Organic Chemistry I

LAB MANUAL

REVISED PRINTING

David Modarelli

Yi Pang

Zhexi Li

Kendall Hunt
publishing company

Cover photo © Shutterstock.com

Kendall Hunt
publishing company

www.kendallhunt.com
Send all inquiries to:
4050 Westmark Drive
Dubuque, IA 52004-1840

Revised 2020 Printing

ISBN: 978-1-7924-3824-0

Published in the United States of America

Table of Contents

Experimental Guidelines

1. **Before the experiment:**
   - Familiarize yourself with the experiment by reading the background information and experimental procedures for the lab in this manual.
   - You are required to bring a calculator to every discussion.
2. **During the experiment:**
   - Experiments can only be completed within your assigned laboratory period.
   - Do not miss the first lab period of any experiment because the starting materials will be removed by the second lab period.
   - Show up for lab on time. Tardiness will prevent you from obtaining any necessary equipment and/or reagents.
   - If you will be late or absent, please inform your instructor and the stockroom manager, Dr. Joanne Li. To ensure full credit on your lab, you must bring a valid reason for your absence as well (i.e., doctor's note, auto receipt, etc.). Skipping lab because you were tired or had something else to do is not a valid excuse.
   - If you do NOT show up for the first lab period in a sequence without letting your instructor and stockroom know, you will NOT be allowed to start the lab during the second period without their approval.
   - You are required to have a laboratory notebook in the form of a bound composition notebook.
   - Laboratory notebooks must be completed in ink and the pages numbered.
   - The lab notebook is just that, a notebook, any material labeled as observations must be done during the laboratory period.
   - Your Teaching Assistant must sign and date your laboratory notebook after every lab period you attend for credit on the experiment.
3. **After the experiment:**
   - Laboratory notebooks will be collected as indicated by your Teaching Assistants and graded according to the requirements given in the handout. Neatness counts!!!
   - Products will be turned in at the completion of an experiment. They may take the form of an actual chemical compound (solid or liquid), a graph or other interpretation of data, or a set of calculations. All chemical products will be checked for authenticity.
   - Your Teaching Assistants reserve the right to deduct points for failure to clean your work areas, including the balance area and the melting point apparatuses.

Laboratory Equipment

If you drop the course, remember that you must check out of the lab. It will not be done "automatically". If you do not check out properly, a $5.00 clean-up fee will be assessed which can prevent you from registration and graduation.

Breakage has been incorporated into the course fee. You will be billed for any breakage over $15.

Safety

No shorts, sandals (even with socks), tube tops, half shirts or any other attire that exposes the torso, legs, or feet is permitted in the labs. These regulations are state and federal laws.

Goggles must be worn (over the eyes, not the forehead or neck) at all time in the labs. The first time you do not have goggles on you will receive a warning, the second time you will lose 5 points, after that you will be asked to end your experiment for that day.

If you are pregnant, you may not work in the labs and must drop the course.

No food or beverages are allowed in the labs at any time.

You will not be permitted into the teaching labs without goggles and proper attire. NO EXCEPTIONS, as this is state law.

Your laboratory equipment and workspace must be clean and in good working order at the end of your experiments. Report any discrepancies to your TA (i.e., nonfunctioning vacuum, lights, leaks, broken glass, etc.)

All injuries, no matter how minor, must be reported to your TA immediately for your protection. Serious injuries must be reported to your instructor and Dr. Joanne Li, the stockroom manager.

Do not dispose of glass (pipettes, broken glass, etc.) in the paper trash containers. If the custodians become injured because of this practice, we will no longer have garbage removal services available in the labs.

Any other behavior that threatens the safety of you or those around you will result in your expulsion from that laboratory period.

Please dispose of all chemicals in the appropriate containers. Disposal of certain organic waste down the drain in some cases is a criminal act.

Do not take your products and/or unknowns home with you. If they are not being collected right away, store them in a safe place in your laboratory drawer.

Your signature on the check-in card indicates that you have read and agree to these safety policies.

Academic Dishonesty

The University of Akron takes very seriously Academic Integrity and has outlined a code of student conduct, which may be found at:

http://www.uakron.edu/ogc/UniversityRules/pdf/41-01.pdf

The relevant sections pertaining to cheating are on pp. 5 and 6. As a prospective teacher, scientist, engineer, or physician, you are expected to behave in a professional manner, including refraining from cheating. Cheating is the actual or attempted practice of fraudulent or deceptive acts to gain an unfair advantage in a grade, whether for yourself or for another student. Cheating includes transmitting or receiving by any and all means information about examination questions or labs.

I will not tolerate cheating or any form of academic misconduct. Penalties for cheating range from a 0 or F on a particular lab, through an F for the course.

Sexual Assault—Title IX

The University of Akron is committed to providing an environment free of all forms of discrimination, including sexual violence and sexual harassment. This includes instances of attempted and/or completed sexual assault, domestic and dating violence, gender-based stalking, and sexual harassment. Additional information, resources, support and the University of Akron protocols for responding to sexual violence are available at uakron.edu/title-ix.

Accessibility

In pursuant to University policy #3359-38-01, The University of Akron recognizes its responsibility for creating an institutional atmosphere in which students with disabilities have the opportunity to be successful. Any student who feels he/she may need an accommodation based on the impact of a disability should contact the Office of Accessibility at 330-972-7928 (v), 330-972-5764 (tdd) or access@uakron.edu. The office is located in Simmons Hall Room 105.

After the student's eligibility for services is determined, his/her instructors will be provided a letter which will outline the student's accommodations.

Simplified Laboratory Notebook Guidelines

The completion of the laboratory notebook is of paramount importance in this course. Regardless of your chosen course of study, employment in that field will require the communication of your day-to-day work in a concise and organized format.

It is very important to be honest. Full credit can still be obtained on the notebook assignment, even if the experiment did not work. One mark of a professional is an individual who can properly document a failure, learn from it, and improve their performance in the future based on this information.

The following are steps to help you prepare your laboratory notebook.

1. *Number all pages throughout the entire notebook in the upper right-hand corner.* Allow the first two pages for the Table of Contents.
2. *Date each new entry.* Each entry in your notebook must have its own date, whether it is prelab information, observations, or the conclusion.
3. *Experiment title*
4. *Table of reagents.* Include all starting compounds, products, solvents, and reagents. You may omit drying agents (sodium sulfate, calcium chloride), common acids and bases (HCl, NaOH, etc.) that are used in the work-up. Be sure to include in tabular form the following information:
   (a) name and structure of compound
   (b) its molecular weight
   (c) melting point (solids), boiling point (liquids)
   (d) density (liquids)
   (e) units for the above entries. Units go in the table headings, not in the column itself.
5. **Main reaction(s).* Write balanced equations with carefully drawn structural formulas. Be sure to include all lone pairs of electrons.
6. **Mechanism for main reaction.* Show the mechanism for the main reaction, if any. Use proper arrow drawing format to show the movement of electrons. Most of the mechanisms will be shown during the discussion.
7. **Adjusted stoichiometry.* Conversion of actual quantities of reagents used to moles. This is done during lab after determining the quantities being used.
8. **Limiting reagent.* Determined from adjusted stoichiometry.
9. *Condensed procedure.* Taken from handout or Lab Manual.
10. *Date and sign your notebook after the prelab.* Get a TA to date and sign it as well.
11. *Observations.* Actual steps you did in Lab. Observations are a written record of what you did in lab and what you observed. Your observations should be short and to the point, but should include enough information so that someone else can reproduce your experiment simply by following your observations. Your observations should include the following:
    (a) color and temperature changes
    (b) glassware used (a drawing of the apparatus is better than words)

(c) exact weights and volumes of all chemical used
(d) heating and cooling times
(e) anything else you feel is pertinent

12. *Date and sign your notebook at the end of each lab period.* Get a TA to date and sign it as well.
13. *Conclusions.* Tell what you learned from the experiment; look at the objectives given at the beginning of each handout. Report the results of your experiment, such as melting point of product and % yield of the reaction. Comment on any reasons why the yield or product purity do not correspond to 100%.
14. *Date and sign your notebook.* At the end of the conclusions you should date and sign your notebook.
15. *Product page.* This is a separate page following the conclusions. It should contain the information asked for in the product section of each experiment.
16. Notebooks that are one day late will receive half-credit. Beyond that they will receive a zero.
17. Check the chalkboards in your lab for any special instructions from the Teaching Assistants.
18. Most writing in the field of chemistry is done in the third person, which uses the chemical, reaction, instrument, analytical technique, etc. as the subject. Recently, it has become more acceptable to use the first person, using "I" or "we", in writing. However, it is difficult to teach when it is correct to use the first person as this varies with the subfield and it can appear arrogant if it is done too often or inappropriately. Therefore, avoid using "I" or "we"in these lab notebooks. By using phrases such as "in this work", you often can avoid using "I" or "we". As your skills as a writer develop, you will learn when it is appropriate to use the first person in your subfield. Note that use of "we" usually is preferred to "I" because chemistry is usually a group effort. Even if you did the work alone, you probably were trained by your research advisor or by graduate students; worked on your advisor's ideas; and used his/her lab, equipment, and chemicals, which were obtained by his/her work. Therefore, it is usually dishonest to use "I".

Parts 1–10 must be completed before the experiment is started.

Part 11 is to be completed during the experiment, as if you are taking notes in a lecture class.

Part 12 should occur as soon as the lab period is over.

Parts 13–15 should be completed shortly after the experiment is complete.

Full versus Modified Lab Experimental Write-Ups

Parts 5–8 (marked with an *) are required for full experimental write-ups, not for modified experimentals. The following chart tells you which experiments require modified or full write-ups.

EXPERIMENT TITLE	MODIFIED	FULL
Distribution Coefficient	*	
Acid/Bases		*(No parts 6–8)
Melting Point	*	
Crystallization	*	
Extraction		*
Fractional Distillation	*	
Methylcyclohexenes		*
Catalytic Hydrogenation		(No mechanism)
Bromobutane		*
Diphenylacetylene		*
Selective Reductions		*
Grignard		*
IR	No prelab required	

EXPERIMENT 1

DISTRIBUTION COEFFICIENTS

Objective

- To introduce the general format of a typical experiment in this course.
- To learn the technique of liquid–liquid extraction to isolate products obtained in an organic experiment.
- To learn the technique of using separatory funnel.

Background

Separatory funnels are commonly used in an organic laboratory. When two insoluble solvents are added into a separatory funnel, solvents tend to separate into two layers, due to their insolubility and different densities. Liquids with higher densities are at the bottom (Figure 1).

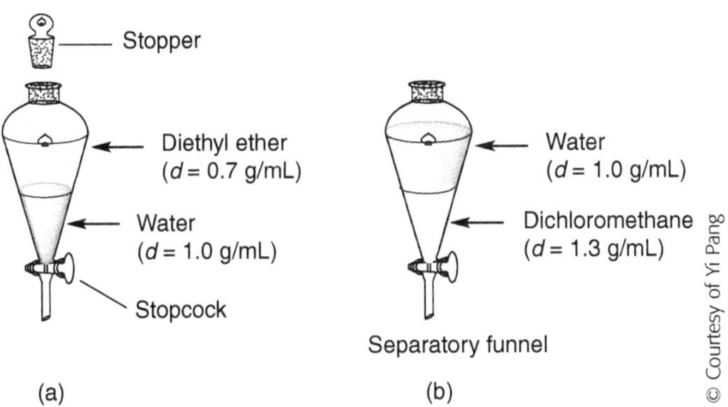

Figure 1. Two solvents in a separatory funnel containing (a) ether and water and (b) water and dichloromethane, which are separated into two layers.

Table 1. Density of common solvents

SOLVENT	DENSITY (g/mL)
Hexane	0.695
Diethyl ether	0.708
Water	1.000
Dichloromethane	1.325

An organic compound often exhibits different solubility in a polar and nonpolar solvent. On the basis of the "*like dissolves like*" principle, the solubility of a solute molecule in a specific solvent is determined by the structural similarity between the solute and the solvent. As an example, phenol includes a hydrophobic and hydrophilic segment (as shown in the structure on right). Due to this structural feature, it has certain solubility in both hydrophobic and hydrophilic solvents. One gram of phenol dissolves in about 15 mL of water (a polar solvent), and 12 mL of benzene (a nonpolar solvent).

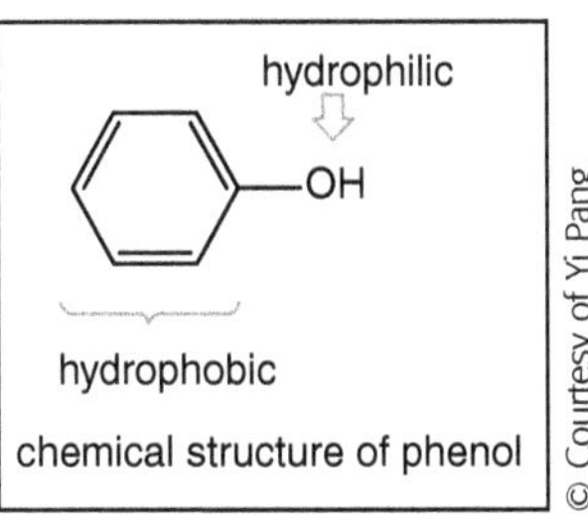

chemical structure of phenol

Solid–liquid extraction. Extractions are one of the oldest chemical operations. Extractions can be classified into solid–liquid and liquid–liquid extraction, depending on the physical properties of the involved materials (for two layers). The preparation of coffee or tea involves the extraction of flavor components from dried solid vegetables. During a laundry process, oil, dirt or fruit juice stains are removed from dirty clothes. These are examples of solid–liquid extraction, where a solvent is used to extract the compounds from the insoluble solid. In a research laboratory, a Soxhlet extractor is often used for solid–liquid extraction, in order to extract soluble materials from the insoluble substrates.

Liquid–liquid extraction, also known as *solvent extraction* and *partitioning*, is a common technique used in the organic laboratory. For example, an organic reaction often gives by-products, some are inorganic and some are organic. Thus, after a reaction is complete, it is necessary to do a *workup*, in order to separate the desired organic product(s) from the mixture of byproducts. Using the water–ether solvent pair for extraction can separate the organic product(s) that is in ether layer, from the inorganic by-product(s) that is(are) in aqueous layer. Separation of organic product(s) from water-soluble by-products is based on their relative solubility in two immiscible liquids. This experiment demonstrates the operation and application of liquid–liquid extraction.

Partition Coefficient

An organic compound (such as phenol) can dissolve in both water and an organic solvent (such as benzene). When the compound is added into an immiscible liquid consisting of two different solvents, the compound will dissolve in both solvents. The equilibrium process is governed by the solubility of the compound in the two solvents. The ratio of solubility is called *partition coefficient k* (or *distribution coefficient*). The *partition coefficient k* is a constant for a given substance, pair of solvents, and temperature.

Let's examine the distribution of a compound **C** (such as 1-butanol), which is slightly soluble in water (6 g/100 mL) but more soluble in ether (12 g/100 mL). The ratio of solubility is:

$$k = \frac{\text{concentration of } \mathbf{C} \text{ in } t\text{-butyl methyl ether}}{\text{concentration of } \mathbf{C} \text{ in water}}$$

$$= \frac{\text{12 g/100 mL ether}}{\text{6 g/100 mL water}} = 2$$

A simple problem: If one has a solution of 6 g of compound **C** in 100 mL of water, and plan to extract **C** with 100 mL of ether. If the "partition coefficient" $k = 2$, how many grams of **C** will be extracted?

In one experiment, 100 mL ether is used in one portion, and shaken with a solution of 6 g of **C** in 100 mL of water. We can assume that $\boldsymbol{x}$ gram of **C** is redistributed in the 100 mL ether layer after shaking. Then,

$$k = \frac{\text{concentration of } \mathbf{C} \text{ in ether}}{\text{concentration of } \mathbf{C} \text{ in water}} = \frac{x \text{ g of } \mathbf{C}\text{/100 mL (in ether)}}{(6-x) \text{ g of } \mathbf{C}\text{/100 mL (in water)}}$$

$$= \frac{x}{6-x} = 2$$

After solving for x, we have $x = 4$.

The result shows that, in a single extraction, 4 g of **C** is distributed in ether layer, and 2 g of **C** in water layer.

It is, however, more efficient to extract the 100 mL of aqueous *twice* with two 50-mL portions of ether (i.e., 50-mL + 50-mL ether), rather than *once* with a single 100-mL portion. This can be shown in the calculation below. When using the *first 50-mL ether* to extract, we have

$$k = \frac{x \text{ g of } \mathbf{C}\text{/50mL (in ether)}}{(6-x) \text{ g of } \mathbf{C}\text{/100 mL (in water)}} = \frac{2x}{6-x} = 2$$

from which,

$x = 3.0$ (indicating that 3.0 g of **C** in ether layer);

$(6 - x) = 3.0$ g of **C** in water layer.

If this 3.0 g/100 mL of water solution is extracted with the *second 50-mL ether*, we have

$$k = \frac{x' \text{ g of } \mathbf{C}/50 \text{ mL (in ether)}}{(3-x') \text{ g of } \mathbf{C}/100 \text{ mL (in water)}} = \frac{2x'}{3-x'} = 2$$

Solving the equation gives us that $x' = 1.5$, showing that 1.5 g of **C** will be in the ether layer and 1.5 g in the water layer. Therefore, extraction with two 50-mL portions will extract $(x + x') = 3.0 \text{ g} + 1.5 \text{ g} = 4.5$ g of compound **C**, whereas one extraction with a single 100-mL portion of ether removes only 4.0 g of **C**. It is generally true that *multiple extractions with smaller amounts of solvent are more efficient than a single extraction with the same total amount of solvent.*

Mixing and Separating the Layer

The liquid–liquid extraction involves mixing and separation of two different layers. *For microscale separation* (e.g., for less than 5 mL liquid), mixing and separation can be accomplished with a pipette with little product loss. Part of the liquid (from either of two layers) can be drawn up with a pipette, and the liquid can be rapidly expelled into the remaining liquid for mixing. Then the layers are allowed to separate and the bottom layer is separated by drawing it up into a pipette and transferring it to a different container.

For macroscale separations, a separatory funnel (Figure 1) is used for mixing and separating the organic layer from the aqueous layer. In macroscale experiments, a frequently used method (for working up) is to dilute the reaction mixture with water and extract the product with an organic solvent, such as ether, in a separatory funnel. In the first step, the stoppered separatory funnel is shaken to distribute the reaction product (or compound) between the immiscible solvents, for example, ether and water. When doing the shaking, the stopper of the funnel is held in place by one hand and the stopcock by the other (Figure 2a). After a brief shake or two, the funnel is held in the inverted position (Figure 2b), and the stopcock is opened cautiously (with the funnel stem pointed away from nearby persons) to release pressure. The mixture can then be shaken more vigorously, with pressure released as necessary. When equilibrium is established, the organic reaction product is mainly distributed into the upper ether layer, while inorganic salts, acids, or bases pass into the water layer. The bottom layer then can be drained into an Erlenmeyer flask. The separatory funnel should be supported in a ring stand (Figure 2c).

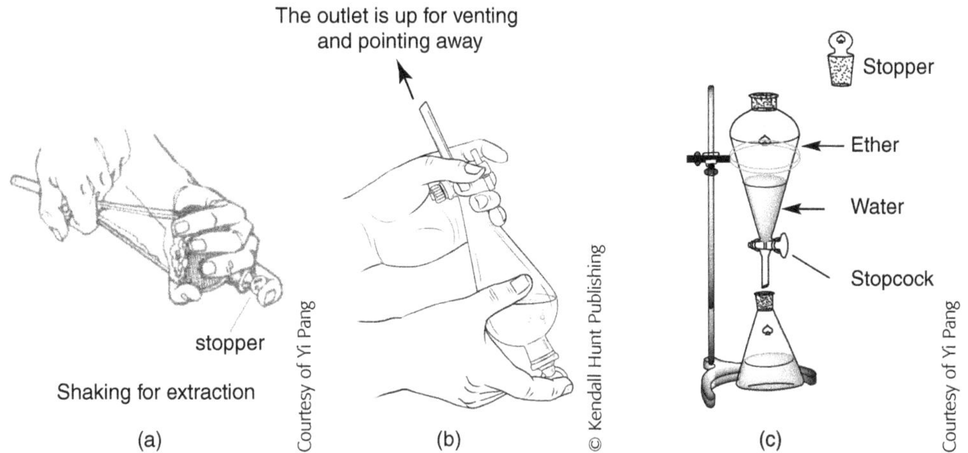

Figure 2. (a) Correct position for holding a separatory funnel when shaking. (b) When venting, always point outlet away from yourself and your neighbors. (c) A separatory funnel is supported on a ring stand, and the bottom layer can be drained into a flask.

(Source: William's textbook, 6 ed., p. 135, Figure 7.3)

Experiment: Crotonic Acid

- Obtain a 25- or 50-mL-graduated cylinder, and a burette from the stockroom.
- Weigh out 90–110 mg of crotonic acid. Record the exact weight.
- Add the acid along with 25 mL of water and 25 mL of methylene chloride to the separatory funnel (make sure the stopcock is closed). Measure the mL as accurately as possible (use graduated cylinder).
- Stopper the separatory funnel, invert it, and carefully release the pressure by opening the stopcock (venting). Close the stopcock and vigorously shake the separatory funnel. Periodically stop the shaking to vent any pressure.

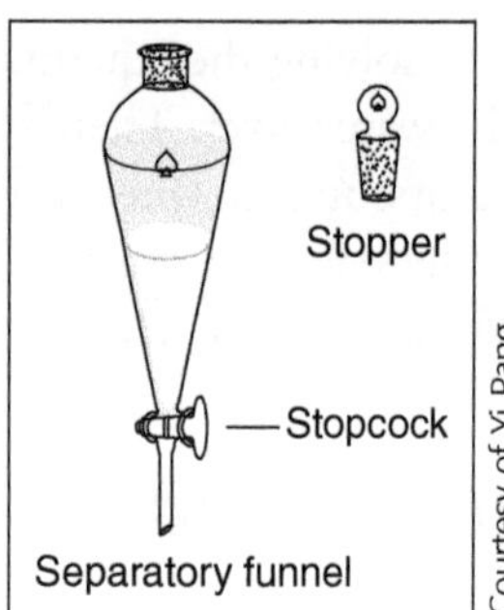

Separatory funnel

© Courtesy of Yi Pang

- Allow the separatory funnel to stand until the two layers are completely separated. Make sure there is not a globule of the lighter liquid caught in the narrow point near the stopcock.
- Which layer is the aqueous layer?
- Remove the stopper and carefully drain off the bottom layer into a 125-mL Erlenmeyer flask (Flask 1).
- Swirl the separatory funnel to coagulate the liquid that adheres to the walls, and make sure that the heavier liquid is completely transferred to the first flask (i.e. Flask 1).
- Pour the remaining liquid through the top of the funnel into the second 125-mL Erlenmeyer flask (Flask 2).
- Rinse the funnel with 2–3 mL of distilled water and transfer this also to Flask 2.
- Finally, drain the liquid in the bore of the stopcock into Flask 1.
- Add four drops of phenolphthalein solution to the aqueous solution (Flask 2) and titrate with 0.10 M NaOH just to a permanent pink color (one that does not fade within 20 s). Do not add too much.
- Record the volume of NaOH.
- Add 10 mL of water to the methylene chloride solution (Flask 1), and four drops of phenolphthalein solution.
- Titrate this solution as above, but swirl vigorously while adding the base. This titration will be more difficult than the first, since for neutralization the crotonic acid will have to go from the organic into the aqueous phase. The pink color will be seen in the aqueous medium before the titration is complete, but will fade as mixing takes place.
- Titrate to a permanent pink color that does not fade within 20 s when swirled vigorously.
- Record the volume of NaOH (remember that the initial reading was not zero).
- Determine the distribution coefficient.

$$K = \frac{\text{Concn}_{\text{org}}}{\text{Concn}_{\text{aq}}} = \frac{\text{mol}_{\text{org}} / \text{vol}_{\text{org}}}{\text{mol}_{\text{aq}} / \text{vol}_{\text{aq}}} = \frac{(\text{vol of base})_{\text{org}}}{(\text{vol of base})_{\text{aq}}}$$

Benzoic Acid

- Add 150 ± 15 mg (record exact weight) of benzoic acid, 25 mL of CH_2Cl_2 (methylene chloride or dichloromethane), and 25 mL of H_2O to the separatory funnel. Make sure lower stopcock is closed before adding anything to the separatory funnel.
- Stopper the funnel and shake the contents for 2–3 min, as in the previous procedure.

- Allow the funnel to stand until the layers separate.
- Carefully draw off the bottom layer (which solvent is it?) and transfer the upper layer into a 125-mL Erlenmeyer flask, (Flask 3); rinse the separatory funnel with 2 or 3 mL of the (upper) solvent (what is it?); and combine with the solution in Flask 3.
- Add indicator and carefully titrate the solution.
- It is not necessary to titrate the other solution.
- Instead, calculate the amount of benzoic acid in the methylene chloride (use the titration data to calculate the weight of benzoic acid, M.W. 122.12, in the aqueous layer, and subtract from the total sample weight).
- Determine K

$$K = \frac{\text{Concn}_{\text{org}}}{\text{Concn}_{\text{aq}}} = \frac{\text{mol}_{\text{org}} / \text{vol}_{\text{org}}}{\text{mol}_{\text{aq}} / \text{vol}_{\text{aq}}} = \frac{\text{wt}_{\text{total}} - \text{wt}_{\text{aq}}}{\text{wt}_{\text{aq}}}$$

$$\text{wt}_{\text{aq}} = \text{vol}_{\text{base}} \times \text{Molarity}_{\text{base}} \times \text{mol wt}_{\text{benzoic acid}}$$

Laboratory Notebook

- Use the format outlined in the syllabus.
- Modified prelab.

Product

- Answer the following questions in your notebook on the pages following the conclusion:
  1. Suppose you have 300. mL of an aqueous solution that contains 30. g of malononitrile. To isolate malononitrile from this aqueous solution, you decide to use ether as the organic solvent in your liquid-liquid extraction. The solubility of malononitrile in ether at room temperature is 20.0 g/100 mL and in water is 13.3 g/100 mL. Consider the following:
     a. How many grams of malononitrile can be isolated with three extractions using 100. mL of ether each time?
     b. How many grams of malononitrile can be isolated with one extraction using 300. mL of ether?
  2. A common mistake when draining the layers from a separatory funnel is forgetting to remove the stopper from the funnel. Why it necessary to remove the stopper in this step?
  3. The distribution coefficient between hexanes and water for an unknown is 7.5. If a solution containing 10. g of this unknown in 100. mL water is extracted with 100. mL of hexanes, how many grams of the unknown can be isolated? If four extractions using 25 mL portions of hexanes were carried out instead, how many grams of the unknown can be isolated? How many mL of hexanes will be required to remove 98.5% of the unknown from this solution in a single extraction?
- Calculations of the distribution coefficients for crotonic acid and benzoic acid in your experiment.

EXPERIMENT 2

ACIDS AND BASES

Objective

- To learn how to take advantage of the acid–base properties of organic compounds for their separation.
- To further illustrate separation of organic compounds into aqueous and organic phases.

Background

A large amount of reactions in organic chemistry are either acid–base reactions, or are reactions involving an acid or a base. It is therefore critical for students to have a solid understanding of acid–base chemistry. In this section, we will cover one definition of acidity and basicity: the Brønsted–Lowry definition.

Acids and Bases: The Brønsted–Lowry Definition

According to the Brønsted–Lowry definition, a Brønsted acid is a proton (H^+) donor, and a Brønsted base is a proton acceptor. For example, in the reaction between ammonia and water, ammonia is the Brønsted base, and water is the Brønsted acid. As illustrated below:

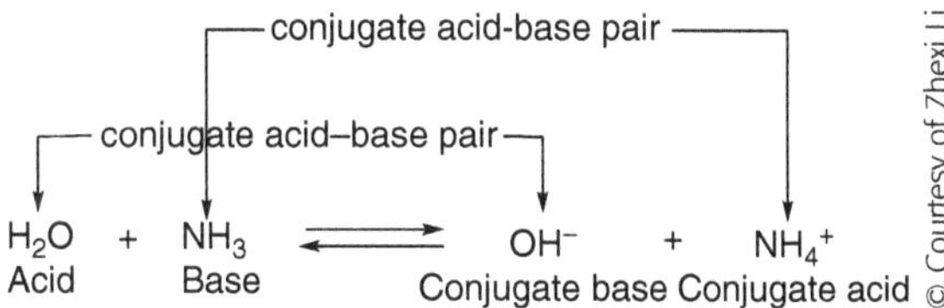

Hydroxide ion, the product formed when H_2O donates a proton, is referred to as the conjugate base. Ammonium ion, the product formed when ammonia accepts a proton, is referred to as the conjugate acid.

The relative strength of an acid in water can be expressed by K_a, the acid dissociation constant. Consider the generic acid–base reaction below:

$$HA + H_2O \rightleftharpoons A^- + H_3O^+$$

The equilibrium constant $K_{eq} = \dfrac{[H_3O^+][A^-]}{[HA][H_2O]}$

The acid dissociation constant $K_a = K_{eq}[H_2O] = \dfrac{[H_3O^+][A^-]}{[HA]}$

The strength of an acid is usually expressed in terms of pK_a rather than K_a, where:

$$pK_a = -\log K_a$$

The pK_a values of some common acids are listed in Table 1. Note the following trends:
The **smaller the pK_a** value the **stronger** the acid, and vice versa.
The **stronger** the acid the **weaker** and more stable its conjugate base, and vice versa.

Table 1. pK_a Values of Some Common Acids

WEAKER ACID	ACID	pK_a	CONJUGATE BASE	STRONGER BASE
↓	H_2O	15.7	OH^-	↑
	C_6H_5OH	9.95	$C_6H_5O^-$	
	NH_4^+	9.24	NH_3	
	H_2CO_3	6.35	HCO_3^-	
	C_6H_5COOH	4.17	$C_6H_5COO^-$	
Stronger acid	HCl	−7	Cl^-	Weaker base

Separation of Organic Compounds by Acid–Base Extractions

The liquid–liquid extraction technique takes advantage of the "*like dissolves like*" principle to separate compounds based on their relative solubility in a two-solvent system. However, this can be problematic if the compounds to be separated are all organic compounds. Most organic compounds have higher solubility in organic solvents than in water. Luckily, this problem can be circumvented by carrying out a simple acid–base reaction to convert one of the compounds to its ionic form, which will be more soluble in water than in organic solvents. After separation, the ionic form can then be converted back to the original form with another acid–base reaction.

This technique is particularly useful for isolating amines (weak bases), phenols (weak acids), and carboxylic acids (strong acids) from a mixture of organic compounds. These molecules can be interconverted between their nonionic forms and ionic forms simply by adjusting the pH of solution (Figure 1).

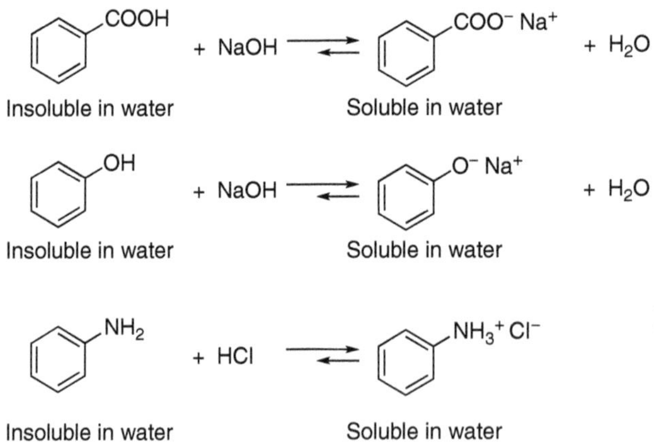

Figure 1. Acid–base reactions of benzoic acid, phenol, and aniline.

Using this strategy, we can separate a mixture of organic compounds by a series of acid–base extractions (Figure 2).

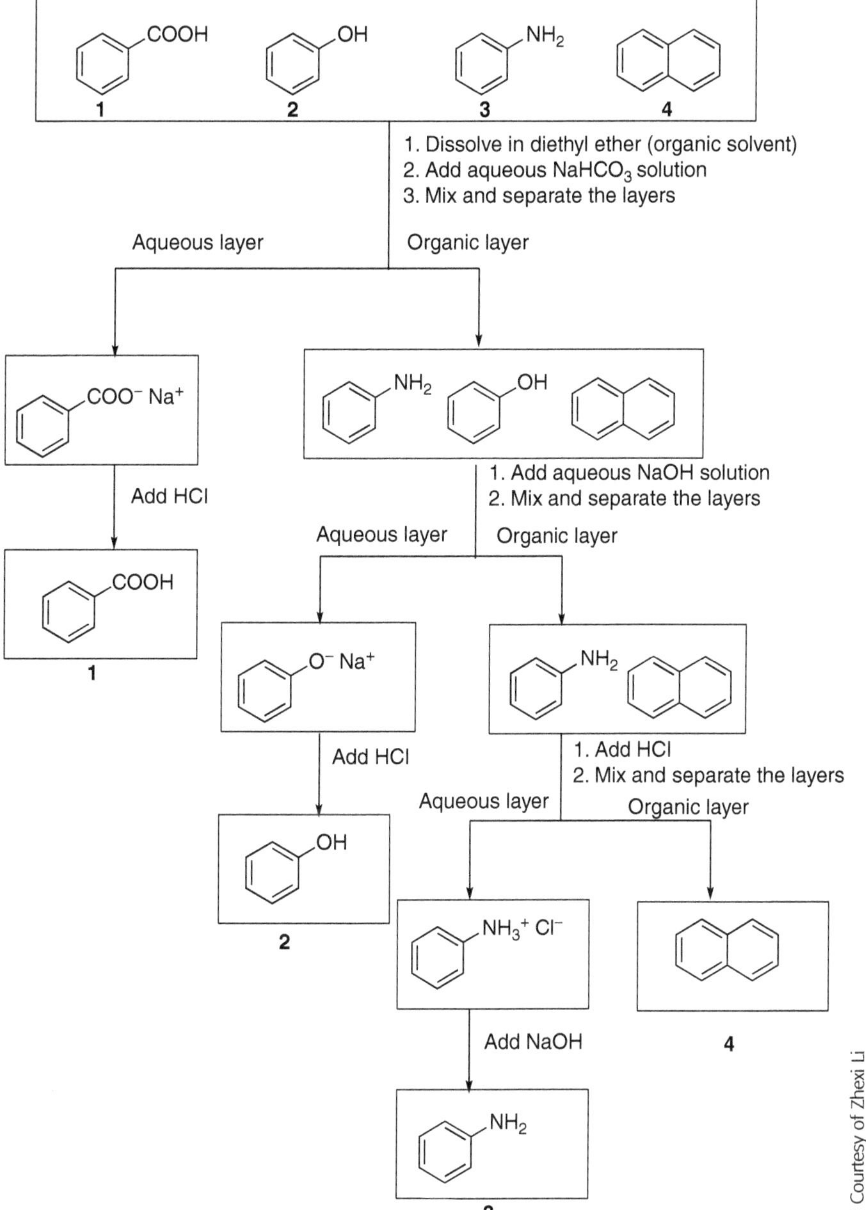

Figure 2. A flow chart for the separation of benzoic acid, phenol, aniline, and naphthalene by acid–base extractions.

In the flow chart above, benzoic acid (a strong acid) was selectively deprotonated by using a weak base ($NaHCO_3$) while the phenol (a weak acid) stayed protonated. This is possible because of their respective acid strengths:

The acid–base reaction between benzoic acid and sodium bicarbonate:

$$C_6H_5COOH + HCO_3^- \rightleftharpoons C_6H_5COO^- + H_2CO_3$$

$pK_a = 4.17$	$pK_a = 6.35$
Stronger acid	Weaker acid

The equilibrium constant K_{eq} for this reaction is:

$$pK_{eq} = 4.17 - 6.35 = -2.18$$
$$K_{eq} = 1.51 \times 10^2$$

Since $K_{eq} > 1$, the reaction equilibrium lies to the right. This is because benzoic acid can be converted to sodium benzoate by reaction with $NaHCO_3$.

The acid–base reaction between phenol and sodium bicarbonate is shown below:

$$C_6H_5OH + HCO_3^- \rightleftharpoons C_6H_5O^- + H_2CO_3$$

$pK_a = 9.95$	$pK_a = 6.35$
Weaker acid	Stronger acid

The equilibrium constant K_{eq} for this reaction is:

$$pK_{eq} = 9.95 - 6.35 = 3.60$$
$$K_{eq} = 2.51 \times 10^{-4}$$

Since $K_{eq} < 1$, the reaction equilibrium lies to the left. To convert phenol to the phenoxide ion, a stronger base (NaOH) will have to be used.

Experiment

- Weigh out approximately 0.4 g of benzoic acid, *m*-nitroaniline, and β-naphthol [2-napthol] and record the exact weights of each. Note your observations on the appearance of each of the three compounds.
- Divide each sample into five approximately equal portions on a watch glass with a spatula.

Benzoic Acid

- Transfer the 80 mg samples of benzoic acid to five 4-inch test tubes.
- To the first add 1 mL of water, to the second add 1 mL of ether (diethylether), to the third add 1 mL of 1 M NaOH, and to the fourth add l mL of 1 M HCl.
- Mix the contents of each tube thoroughly.
- Record the results for each tube:

  1. Did the benzoic acid dissolve in water?
  2. Did it dissolve in ether?
  3. If you had added 1 mL of water and 1 mL of ether, where would the benzoic acid be (which layer)?
  4. Did it dissolve in NaOH? Why, when it is insoluble in water?

5. If you had added 1 mL of ether with 1 mL of 1 M NaOH, where would the benzoic acid be (which layer)? In what form (draw the structure)?

6. Did it dissolve in HCl?

7. If you had added 1 mL of 1 M HCl and 1 mL of ether, where would the benzoic acid be? In what form (draw the structure).

- To the test tube with the ether, add 1 mL of 1 M NaOH and carefully shake.

8. Where is the benzoic acid?

- Carefully draw off the aqueous layer (which one?) with a Pasteur pipette and add it to 1 mL of 2 M HCl. What happens?
- Weigh a watch glass, transfer the ether solution to it, and let the ether completely evaporate in your hood.
- What and how much is left?
- Write the equations for the reactions of benzoic acid with aqueous acid and aqueous base, and indicate the relative solubility in each medium.
- To the fifth test tube, add 2 mL of water. Warm the test tube in the steam bath. What happens?
- Reread question #1 and explain the difference.
- Set the hot test tube aside (use a 50 mL beaker as a holder) and let it stand. What happens?
- Save this sample for later; clean the other test tubes for the next tests.

m-Nitroaniline

- Take the sample of *m*-nitroaniline and divide it into five equal portions.
- Treat one portion with water, one with 1 M NaOH, one with 1 M HCl, one with water and sodium bicarbonate, and one with ether.
- Answer questions 1–5 (in the benzoic acid section above) for *m*-nitroaniline.

9. Did it dissolve in the HCl? If you had added 1 mL of 1 M HCl and 1 mL of ether, where would the *m*-nitroaniline be? In what form, draw the structure.

- To the test tube with the ether, add 1-mL of 1 M HCl and (carefully) shake.

10. Where is the *m*-nitroaniline?

- Carefully draw off the aqueous layer and add it to 1 mL of 2 M NaOH. What happens?
- Evaporate the ether solution as before. What and how much is left?
- Write the equations for the reactions of *m*-nitroaniline with aqueous acid and aqueous base, and indicate the relative solubility in each.
- To the fifth test tube, add 2 mL of water and warm in the steam bath. What happens?
- Explain the difference from question #1.
- Set the test tube aside to cool. What happens?

β-Naphthol (a Phenol)

- Divide the sample of β-naphthol into five equal portions.
- Treat one portion with water, one with 1 M NaOH, one with 1 M HCl, one with water and sodium bicarbonate, and one with ether.
- Answer questions 1–5 for β-naphthol.

- To the ether solution, add 1 mL of 1 M NaOH, mix well, separate the layers, and evaporate the ether. What remains and how much?
- Acidify the aqueous portion. What happens?
- Heat the sample that you treated with water. What happens?
- Take the benzoic acid/water sample from before and treat it with a small amount (~half of a microspatula) of sodium bicarbonate.
- What happens? Write an equation.
- Transfer this solution to your separatory funnel and add 25 mL of water and 3 mL of 1 M HCl solution. Check the pH to make sure it is acidic.
- What happens? Write an equation.

Laboratory Notebook

- As per the format in the syllabus.
- Full prelab (no parts 6–8)
- Make sure you make note of the appearance of everything you observed throughout the procedure above. You will need to show the equations for the reactions of benzoic acid, *m*-nitroaniline, and β-naphthol with HCl, NaOH, and $NaHCO_3$.

Product

- On the page following the conclusions, answer the questions asked in the procedure.
- Complete the flow sheet on the next page. Fill the in squares with the correct structure(s) (no formulas). If a layer has nothing but solvent, leave it blank.
- Write equations, using correct structures, for the reactions of benzoic acid, *m*-nitroaniline, and β-naphthol with HCl, NaOH, and $NaHCO_3$. There should be a total of nine equations.

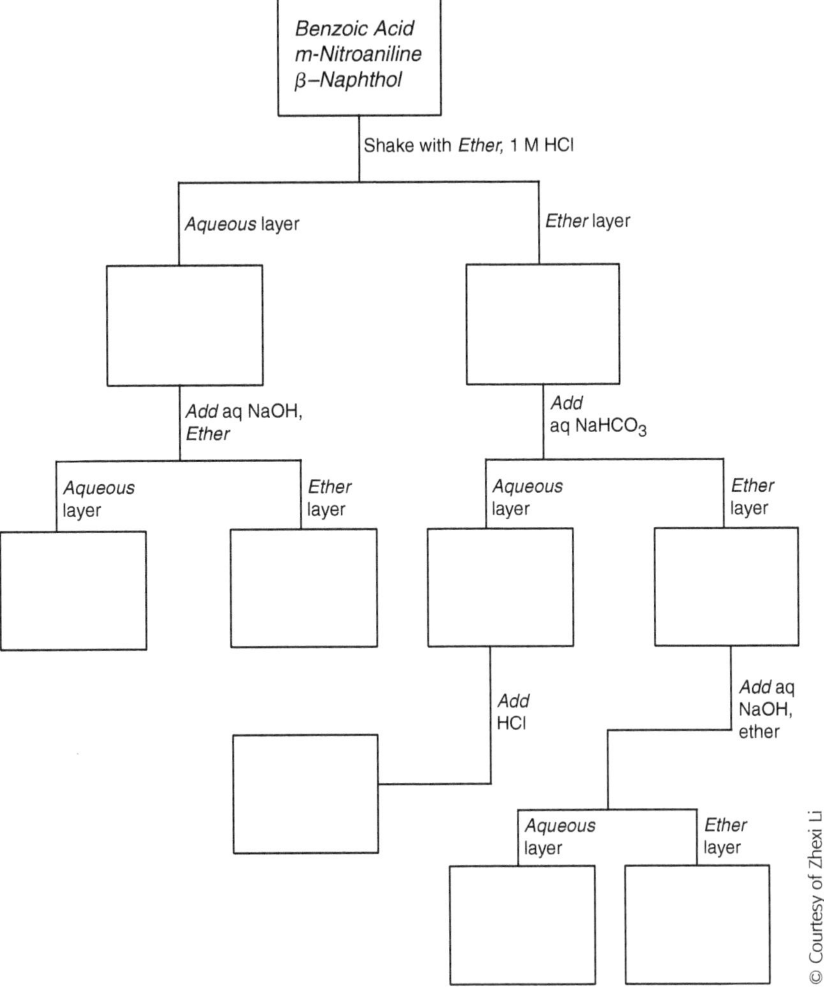
Benzoic Acid
m-Nitroaniline
β–Naphthol
Shake with Ether, 1 M HCl
Aqueous layer
Ether layer
Add aq NaOH, Ether
Add aq $NaHCO_3$
Aqueous layer
Ether layer
Aqueous layer
Ether layer
Add HCl
Add aq NaOH, ether
Aqueous layer
Ether layer
© Courtesy of Zhexi Li

EXPERIMENT 3

MELTING POINTS

[People in Room 410 will do experiment 3 first, those in Room 412 will do experiment 4 first]

Objective

- To learn how to identify/purify an organic compound.
- To learn how impurities affect the melting point of a pure material.
- To learn the technique of using the melting point apparatus.

Background

Intermolecular forces (forces *between* two molecules) largely determine the physical properties of a given molecule, in particular, the melting point, boiling point, and solubility. For example, the melting point of a molecule is defined as the amount of energy required to disrupt the intermolecular forces between two molecules. It is important to note that the molecular property that caused these forces still exists, but that the increase in temperature provides enough energy to disrupt the forces, resulting in an increase in the disorder of the solid, causing a phase change in the sample (solid to liquid, or in the case of sublimation, solid to gas). The boiling point, in turn, represents the amount of energy required to disrupt these interactions such that they disappear almost completely and the molecules undergo a phase change from a liquid to a gas. The following introduction will briefly introduce these concepts.

I. *London Dispersion Forces or van der Waals Forces* (~1 kcal mol^{-1})
London Dispersion Forces (LDF) are based on the attractive forces between two adjacent molecules that occur as a result of *temporary dipole* (see next section) moments that arise in every molecule. These interactions are largely based on the size of the molecule and are present in every molecule. In Figure 1, space-filling models of methane (bp −162°C), butane (bp 69 °C), and dodecane (bp 216 °C) are shown for comparison. The space-filling models also show how the intermolecular forces differ among the three compounds, based on the available surface area for interactions between electrons on each compound.

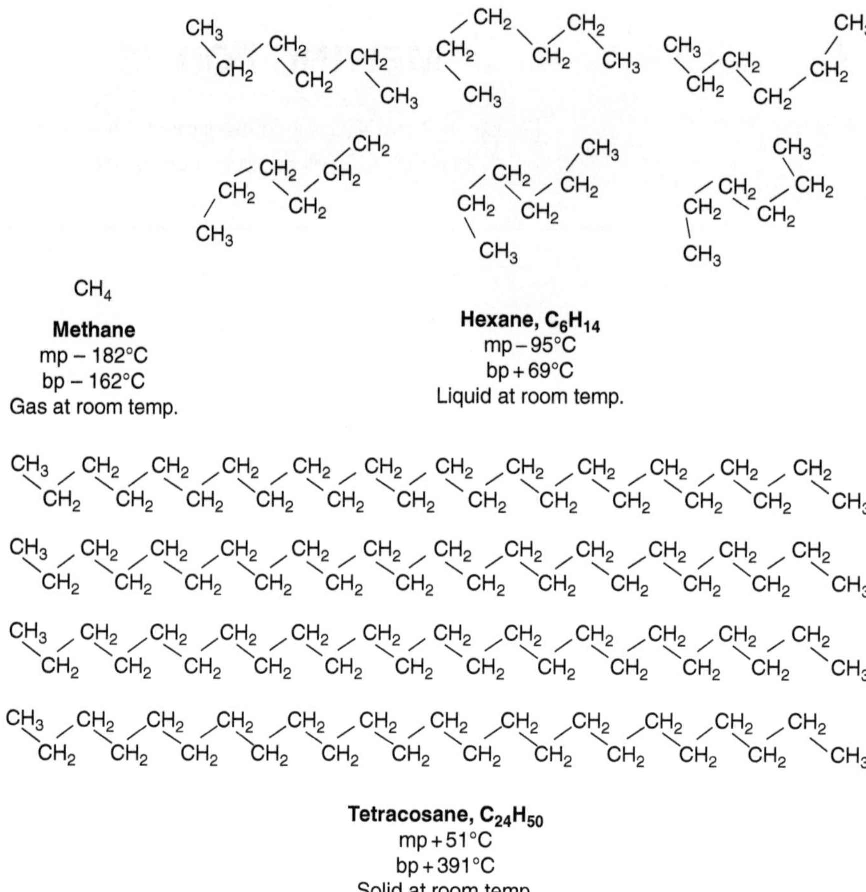

Figure 1. London dispersion forces in methane, hexane and tetracosane.

II. *Dipole–Dipole Interactions* (0.5–3 kcal mol^{-1})

Dipole–dipole interactions are stronger forces than LDFs and are based on the inherent dipole moments present in most non-hydrocarbon molecules. The dipole moments result from the differences in electronegativities between atoms, and so the greater the electronegativity difference, the larger the dipole moment. Dipole–dipole interactions arise when the molecular dipoles of two (or more) molecules line up so that the partial positive charge of one molecule and the partial negative charge of another molecule are attracted to one another (Figure 2). This attraction leads to an increase in order, which results in an increase in melting/boiling point, since the attractive energy (the dipole–dipole interaction!) becomes harder to overcome.

(a)

(b)

Figure 2. Dipole–dipole interactions between two acetone molecules (a) and two methyl chloride molecules (b).

III. *Hydrogen Bonding* (1–12 kcal mol^{-1})

Hydrogen-bonding interactions are strong intermolecular forces that are based on interactions between hydrogens bonded to nitrogen or oxygen (hydrogen-bond donors such as: alcohols, carboxylic acids, amides, and amines) with other atoms having lone pair electrons (hydrogen-bond acceptors such as: ethers, alcohols, ketones/aldehydes, carboxylic acids, amides, and amines). Hydrogen-bonding interactions are particularly important in aqueous (water based) solvents and biological systems, and are involved in protein folding, DNA, etc.

IV. *Ionic Interactions*

Ionic bonds are formed between ions with different charges, such as Na^+ and Cl^- in table salt (NaCl). Some organic molecules, in particular those involved in acid–base reactions, can have attractive forces resulting from ionic interactions. Unlike NaCl, which is a solid, these intermolecular forces often result in soluble organic molecules that are strongly attracted to one another but may or may not be formally considered ionic bonds. Three examples are shown below:

1. Carboxylic acids: The reaction of a carboxylic acid such as acetic acid with sodium hydroxide yields sodium acetate (containing an ionic bond between oxygen and sodium), a solid with a melting point over 300 °C.

$$H_3C{-}C(=O){-}OH + NaOH \rightleftharpoons H_3C{-}C(=O){-}O^-Na^+ + H_2O$$

$pK_a \sim 4.76$ $\qquad$ $pK_a \sim 15.7$

2. Phenols: Phenols are relatively acidic molecules with pK_a values in the 3–12 range. These molecules will usually react with a reasonably strong base such as NaOH in a reaction that results in formation of a salt such as sodium phenoxide (mp 382°C).

$$C_6H_5OH + NaOH \rightleftharpoons C_6H_5O^-Na^+ + H_2O$$

$pK_a \sim 10.0$ $\qquad$ $pK_a \sim 15.7$

3. Amines: Amines are fairly basic (and nucleophilic) molecules that react rapidly with acids to form quaternary ammonium salts. These salts will often be solids at room temperature.

$$C_5H_5N + HCl \rightleftharpoons C_5H_5N^+{-}H\ Cl^-$$

pK_a –8.0 $\qquad$ pK_a ~5.2

V. *Melting Points*

The melting point of a solid is a physical property of that molecule. If it is pure, the sample will melt reproducibly over a very narrow temperature range (usually ~1°C). Because of this reproducibility and narrow temperature range for most samples, melting points are often used to verify sample purity, characterize a newly synthesized molecule, or identify a previously prepared sample.

Organic compounds are usually found as crystals or disordered solids. As the sample is heated, the individual molecules within the solid begin to vibrate, and as the temperature increases, the increased energy becomes great enough to overcome the intermolecular forces described earlier. These vibrations are converted into translational motion, converting the solid into a liquid.

In general, larger molecules will melt at higher temperatures than smaller molecules because they will have more intermolecular interactions. When molecules have similar molecular weights, other factors such as symmetry (higher symmetry generally results in a higher melting point), enantiopurity (racemic mixtures melt at different temperatures than pure *R* or *S* solids), and whether or not the molecule can hydrogen-bond or not.

The melting point of a pure molecule is generally higher than that of an impure compound. In this lab, we will explore the idea of compound purity using mixtures of two compounds in different ratios (mole fractions). Figure 3 shows an example of such an experiment. In this plot, the melting point of two molecules (shown as Molecule X and Molecule Y in Figure 3) having similar melting points is shown as the *y*-axis and the percent of each compound is shown as the *x*-axis. At the points where each molecule is the sole component (i.e., 100% X or 100% Y), the melting point is the greatest. As the mixture becomes closer to 50:50 ratio, the melting point decreases to the so-called *eutectic point*. The reason for the change in melting point is the disruption of the intermolecular forces present in either pure X or pure Y by the impurity (Y or X, respectively).

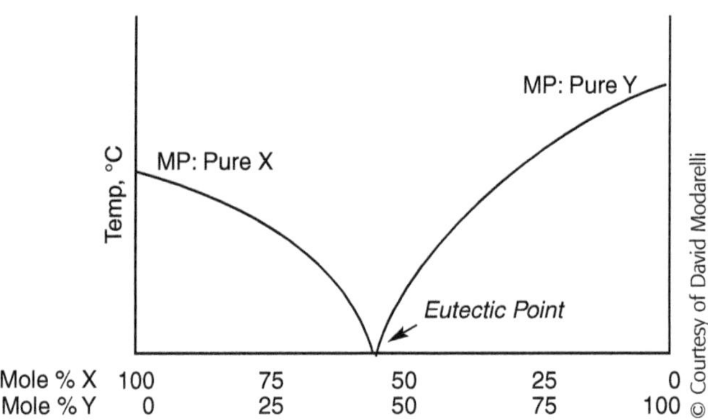

Figure 3. A melting point plot for two compounds (*X* and *Y*). The x-axis shows the percent of each in the mixture, while the y-axis shows the melting point at that particular composition.

In addition to the decreased melting point, the presence of an impurity will increase the range of the melting point, and impure compounds will have ranges well beyond the ~1°C range of a typical pure compound. You should note that an "impurity" can be another solid *or* a liquid (i.e., water, a recrystallization solvent, etc.).

Melting Points

I. *Thermometers*

Several different types of thermometers are available for use in chemical labs. In the melting point lab, we will use a mercury thermometer because of the greater temperature range available (up to ~260 °C). Because mercury is toxic, ***if you break a thermometer***, please **notify your TA or instructor IMMEDIATELY** so they can help you clean it up.

II. *Capillary Tube Sample Holders*

You will use a premade capillary tube for your melting point experiments. These tubes are ~2 mm in diameter and ~4 cm long. To fill one, you will take a small amount of your solid material that is ground into a fine consistency (use a spatula and a watch glass), and place your capillary tube upside down so the open end is against the ground sample and the closed end is pointed up. Press your tube onto your sample, and you should see ~2 to 4 mm of your sample being compressed into the tube. You can then either (a) invert your capillary tube so that the open end faces up, and CAREFULLY strike the closed end hard against the lab benchtop or (b) drop your capillary tube (closed side down!) into the long glass tube in the middle lab. Both techniques will result in your solid sample being transferred from the open end to the closed end of the capillary tube.

III. *Melting Points*

Your melting point will be performed on a small sample, typically <1 mg (the amount you just filled into your capillary tube in the previous section). You can choose to use either one of the two Mel-Temp instruments. One is controlled with a dial, and the other one is fully digital. The analog Mel-Temp apparatus has a mercury thermometer, whereas the digital Mel-Temp displays the temperature on an LCD screen.

If you are using the digital Mel-Temp, please read and follow the instructions posted near the instrument carefully. For the analog Mel-Temp system system, a solid metal block is electrically heated. The heating element is controlled with a dial that varies a transformer—turning the dial to the right increases the amount of heat. Because the Mel-Temp can go MUCH hotter than the mercury thermometer can, it is **VERY IMPORTANT** that you ***turn down and then turn off*** the Mel-Temp instrument when you are finished with your melting point.

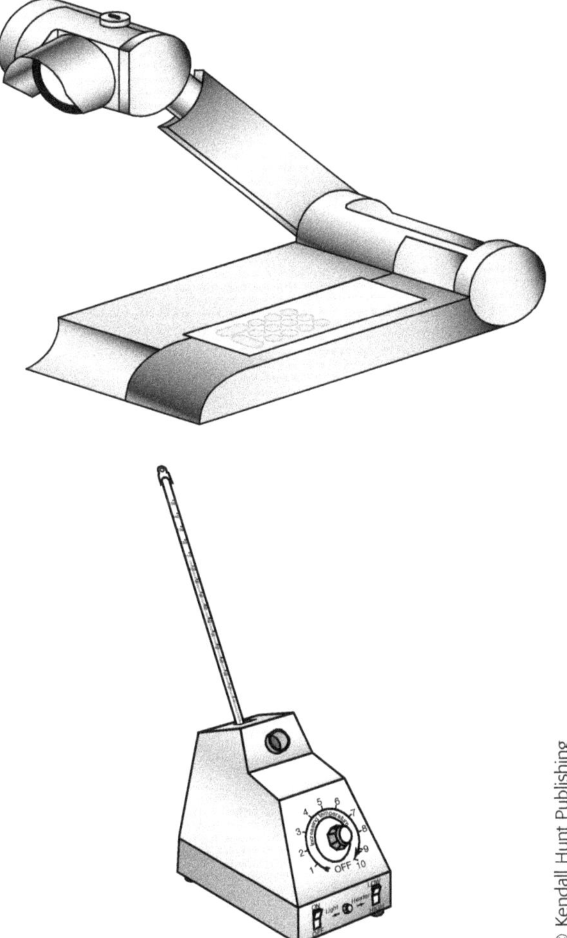

Figure 4. The two different melting point instruments used in this organic lab.

You will determine your melting points in two separate experiments. The first one will involve quickly heating your sample tube in the melting point apparatus. You will watch the sample in the capillary tube through the glass opening in the melting point apparatus as the temperature rises by ~20 °C/min. You should see the sample shrink and then melt as you approach the melting point. Note the temperature at which this process happens. For the second experiment, you will slowly take the melting point apparatus until the temperature is ~20 °C *below* the temperature you determined in your first experiment, and once it reaches this temperature, you will make sure the temperature increases at ~1 °C/min, and then insert your capillary tube and watch it until it shrinks and melts. You will then record this value—this melting point will be the actual melting point of your sample.

Experiment

- Determine the melting point of pure urea and cinnamic acid using the following procedure. Record this data in your notebook.
- Obtain melting point samples of the premade mixtures of cinnamic acid (CA) and urea (90:10, Urea:CA; 50:50, U:CA; and 10:90, U:CA).
- Determine the melting point for each mixture and record this data.
- Construct a plot to show the variation in melting point with % composition.

Determination of an Unknown by Mixed Melting Point

- Obtain an unknown from the stockroom. Determine the melting point.
- Look up the melting point value on the table provided.
- Check the table below and prepare 1:1 mixtures of your unknown with each of the three compounds whose melting points are close to yours using about 5 mg [~one-fourth of a microspatula] of each material on three separate watch glasses.
- Grind and mix the individual mixtures with the microspatula thoroughly.
- Determine the melting point of each mixture.
- Identify your unknown.
- Remember to turn the voltage knob to 0 and turn off the apparatus when you are finished.

Laboratory Notebook

- Use the format outlined in the syllabus.
- Modified pre-lab.

Product

- The melting point and probable identity of your unknown. Be sure to record your unknown #!!!
- A plot of melting points versus % composition for urea–cinnamic acid.

MELTING POINTS OF COMPOUNDS (°C)	
78–81	Acetamide
80–82	Durene
81–83	Vanillin
87–89	*p*-Dibromobenzene

86–88	*p*-Methoxyphenylacetic Acid
88–90	*m*-Dinitrobenzene
100–103	Diphenylacetamide
103–105	*o*-Toluic Acid
104–106	Acetoacetyl *o*-Toluidine
113–114	*p*-Nitrophenol
113–115	Acetanilide
114–116	Fluorene
127–129	Benzoylbenzoic Acid
129–130	2-Furoic Acid
128–129	Benzamide
147–150	Diphenylacetic Acid
152–154	Adipic Acid
152–155	Diphenylthiourea
170–172	Succinic Dihydrazide
171–172	5-Chlorosalicylic Acid
171–173	Isonicotinic Hydrazide

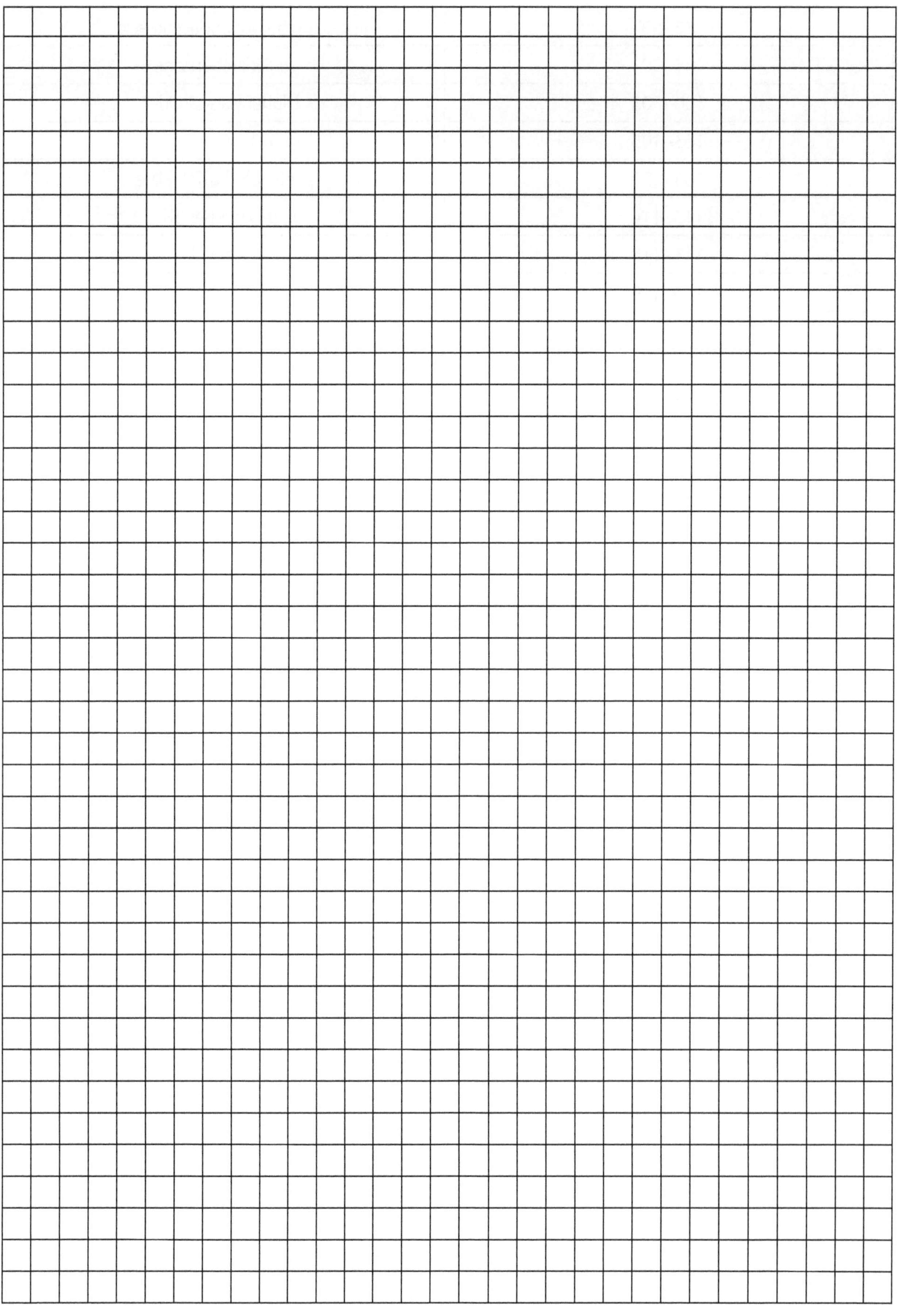

EXPERIMENT 4

CRYSTALLIZATION

[People in Room 410 will do experiment 3 first, those in Room 412 will do experiment 4 first]

Objective

- To introduce the technique of crystallization to purify an organic solid.
- To learn the technique of vacuum filtration to collect a solid material from a supernatant liquid phase.

Background

Recrystallization is a straightforward and powerful method for purifying solid organic compounds. The name recrystallization comes from the process: it involves dissolving a crystalline solid in a hot solvent and let the solid *crystallize again* from the solution. The product obtained through recrystallization usually increase in purity, because the impurities present are generally either removed through filtration or stay dissolved in solution.

Recrystallization is possible because of the fact that the solubility of a solid solute in solvent increases as the temperature of the solvent increases. For example, consider the following scenario: a sample of 5.00 g of compound A is mixed with 0.15 g of compound B (the impurity). The solubility values of compound A and B are tabulated in Table 1.

Table 1. Solubility of Compounds A and B in Water at 100 °C and 25 °C

Compounds	Solubility in 100°C Water	Solubility in 25°C Water
A	15 g per 100 mL	0.15 g per 100 mL
B	11 g per 100 mL	0.25 g per 100 mL

Saturated solution of compound A at 100°C can be made with 33 mL of water.

$$5.00\text{ g} \times \frac{100\text{ mL}}{15\text{ g}} = 33\text{ mL}$$

When this solution is cooled to 25°C, the amount of compound A that stays dissolved is:

$$33\text{ mL} \times \frac{0.15\text{ g}}{100\text{ mL}} = 0.050\text{ g}$$

The remaining 4.95 g will, in theory, crystallize from solution. The impurity (compound B) will stay dissolved in solution at 25°C and be filtered away in the filtrate. Therefore, the maximum percent recovery for this recrystallization is:

$$\frac{4.95\text{ g}}{5.00\text{ g}} \times 100 = 99.0\%$$

Recrystallization usually involves the following steps outlined below:

Step 1. *Choosing the Appropriate Solvent*

Keeping in mind the principle "*like dissolves like*", it is easy to discern that water or alcoholic solvents will dissolve polar compounds and nonpolar solvents such as hexanes or benzene will dissolve nonpolar compounds. An ideal solvent should fulfill the following criteria:

- The desired compound has high solubility in the solvent when the solution is hot, and low solubility when the solution is cold.
- The desired compound will not react with the recrystallizing solvent.
- The solvent is safe to handle, inexpensive, and can be easily removed from the crystallized product.

Some common solvents are listed in Table 2, in the order of decreasing polarity.

Table 2. Common Solvents for Recrystallization

Solvent	bp (°C)	Remarks
Water	100	Usually the solvent of choice. Cheap, non-toxic, nonflammable. However, it is not easily removed from crystals
Methanol	64	Flammable. Can be easily removed from crystals
Ethanol	78	Flammable. Can be easily removed from crystals
Ethyl acetate	78	Flammable. Can be easily removed from crystals
Dichloromethane	40	Nonflammable. Low boiling point
Toluene	111	Flammable. High boiling point. Difficult to remove from crystals (Note: in practice, avoid use toluene due to its high b.p.)
Hexane	69	Flammable. Can be easily removed from crystals.

Sometimes it is necessary to use mixed solvents for recrystallization. This is usually necessary when the compound is too soluble in one solvent and only sparingly soluble in another solvent. The solvents used in a mixed solvent pair must be miscible with each other. When using mixed solvents, the compound is dissolved in a small amount of the better solvent nearing the boiling point, and then the poorer solvent is slowly added until cloudiness is observed. Crystals should form in this mixture upon cooling. Some examples of solvent pairs are ethanol–water, toluene–hexanes, and acetic acid–water.

Step 2. *Dissolving the Impure Sample*

The impure sample is placed in a test tube with a small amount of recrystallization solvent. The mixture is heated on a steam bath (or a sand bath) to a boil. If the sample is not completely dissolved, slowly add more solvent to the tube, keeping the mixture at its boiling point. Once all solid has been dissolved, remove the tube from heat and allow it to cool to room temperature.

Step 3. *Removing Colored Contaminants Using Activated Charcoal*

If the solution is heavily colored when the compound is known to be colorless or lightly colored, a small amount of intensely colored impurities might be present. These impurities can be removed by adding activated charcoal.

A small amount of activated charcoal (~0.1 g, or the size of a match head) is added to the colored solution, keep the solution hot to prevent any crystallization of the product. Let the mixture settle. If the color disappears, remove the activated charcoal by filtration. If the color persists, add more activated charcoal and repeat the process. Since activated charcoal is a fine powder, it is necessary to use a small bed of Celite® or other filter aid for the filtration. Be cautious with the amount of activated charcoal being added. Adding too much activated charcoal will result in a lower yield as the product can be absorbed onto the charcoal as well.

Step 4. *Filtering the Solution*

Any insoluble solids must be removed from the solution. This is usually done through filtration or using a Pasteur pipette. Gravity filtration is often used instead of vacuum filtration to prevent premature crystallization. We will use a Pasteur pipette to withdraw solvent from the solution (Figure 1).

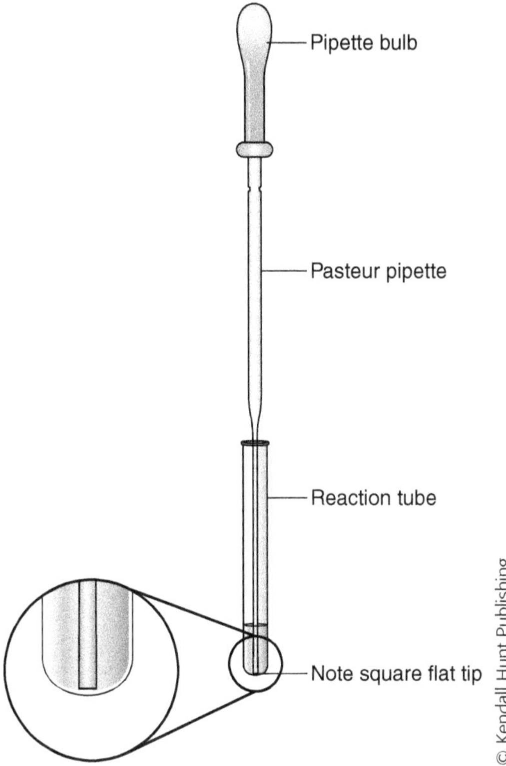

Figure 1. Withdrawal of supernatant using a Pasteur pipette.

Step 5. *Recrystallization*

This step will start with a saturated hot solution of the desired compound. If more solvent has been added to the solution to facilitate filtration, the excess solvent will have to be removed by boiling or a stream of air. Once a hot, saturated solution is obtained, allow it to cool slowly to room temperature. The rate of cooling determines the size and purity of the crystallized product. If the solution is cooled too quickly, smaller crystals will form along with impurities. The smaller crystals can also make filtering and drying more difficult. A slower rate of cooling can be achieved by placing the test tube in a poor conductor of heat, like cotton or paper towels. Once the test tube is cooled to room temperature, put it in an ice bath to facilitate further crystallization.

If the formation of crystals is not observed upon cooling, the solution might be supersaturated. Either adding a seed crystal or scratching the inside of the test tube, just below the liquid surface, can induce crystallization of a supersaturated solution.

Step 6. *Collecting and Washing the Crystals*

The crystals formed will need to be removed from the mother liquor by filtration (Figure 2). Alternatively, a Pasteur pipette can be used to withdraw the mother liquor from crystals (Figure 1). A small amount of ice-cold recrystallization solvent will need to be used to remove any impurities adhered to the surface of crystals.

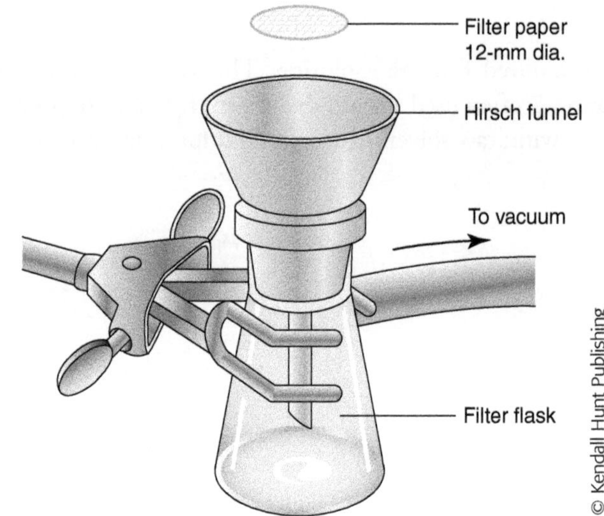

Figure 2. Filtration using a Hirsch funnel

Step 7. *Drying the Crystals*

The crystals can be either dried in the test tube under reduced pressure (see Figure 3) or air dried on the Hirsch funnel by leaving the vacuum on after the last wash.

Experiment

Phthalic Acid Crystallization from Water

- Weigh out 60 mg of phthalic acid and place into a clean 4 in. test tube.
- Add the minimum volume of water needed to cover the sample.
- Using your steam bath gently heat the solution to boiling. As soon as boiling begins, continue to add water dropwise until the entire solid just dissolves.
- Remove the test tube from the steam bath and place in a cooling rack. You should begin to observe recrystallization of the sample upon cooling to ~25°C. If no crystals are visible, scratch the sides of the test tube using a microspatula to induce crystallization.
- After the tube reaches ~25°C, cool in a small beaker of ice water for a few minutes.
- Expel the air from a Pasteur pipette and then place the tip near the bottom of the test tube. Gently raise the syringe nozzle while withdrawing the water from the tube, leaving the solid crystals behind (Figure 1).
- Cool the tube in ice bath and add a few drops of ice-cold ethanol to the tube to remove water from the semidry crystals. Quickly pipette away the ethanol, as it will dissolve your sample.
- Attach the test tube to a vacuum pump. The crystals are dry when the solid is easily tapped off the sides of the test tube (Figure 3).
- Scrape the crystals out of the test tube using a microspatula. Obtain the weight of your recrystallized sample and calculate the percent recovery/yield.
- Obtain the before and after melting point of your sample and place in a sample bag to give to your TA.

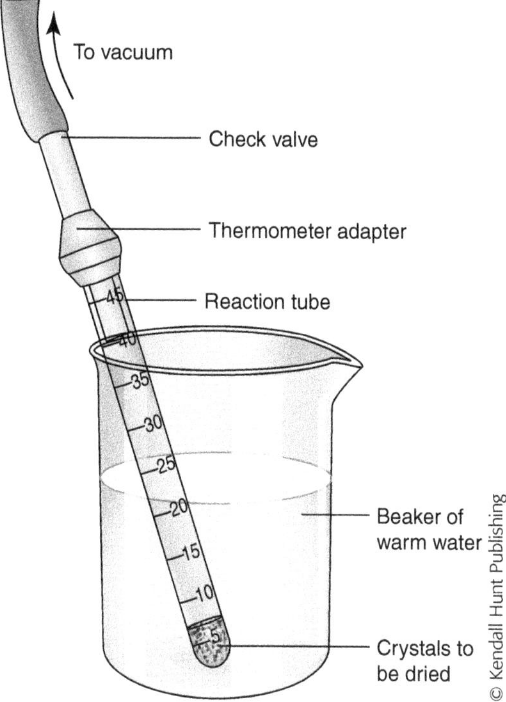

Figure 3. Evaporation of the solvent under reduced pressure.

Benzoic Acid Crystallization from Cyclohexane

- Weigh out 60 mg of benzoic acid and place into a clean 4 in. test tube.
- Add the minimum volume of cyclohexane needed to cover the sample.
- Using your steam bath gently heat the solution to boiling. As soon as boiling begins, add cyclohexane (only if needed) dropwise until the entire solid just dissolves.
- Remove the test tube from the steam bath and place in a cooling rack. You should begin to observe recrystallization upon cooling to ~25°C. If no crystals are visible, scratch the sides of the test tube using a microspatula to induce crystallization.
- After the tube reaches ~25°C, cool in a small beaker of ice water for a few minutes.
- Expel the air from a P.asteur pipette, and then place the tip near the bottom of the test tube. Gently raise the syringe nozzle while withdrawing the organic solvent from the tube, leaving the solid crystals behind.
- Cool the tube in ice and add a few drops of ice-cold cyclohexanes and quickly pipette away the solution (this process is called "washing your sample," in order to remove impurities from the crystals).
- Attach the test tube to a vacuum pump. The crystals are dry when the solid is easily tapped off the sides of the test tube.
- Scrape crystals out of the test tube using a microspatula. Obtain the weight of your sample and calculate the percent recovery/yield.
- Collect your sample (before and after taking the melting point) and place in a sample bag to give to your TA.

Decolorizing a Solution with Charcoal

- Into a clean 4 in. test tube, place 1.0 mL of a solution of methylene blue dye.
- To the tube add a small amount of charcoal (the size of a match head).
- *Gently* heat the contents of the test tube to boiling and observe the color by holding the tube in front of a piece of white paper.
- For clean up: Set up your vacuum filtration and add filter paper to your Hirsch funnel while the vacuum is ON, and add a scoop of celite to the top of the filter paper. Shake and pour your solution into the funnel. All solid waste should be disposed of in the solid waste containers at the front of the room.

Purification of an Unknown

- Obtain an unknown from the stockroom. **After** finding a *suitable solvent,* recrystallize 200 mg of unknown and collect the crystals using the Hirsch funnel. There should be no need to decolorize. Follow the seven steps outlined in the background section.
- Obtain the weight of your product/unknown and calculate the percent yield.
- Obtain the before and after melting point of your sample and place in a sample bag to give to your TA.

Laboratory Notebook

- Use the format outlined in the syllabus.
- Modified prelab.

Product

- To be turned in at the end of the experiment, the three purified solid materials from the experiment: phthalic acid, benzoic acid, and the unknown. They must be labeled as instructed by your TA.
- Calculations for percent recovery for each compound.
- The "before" and "after" melting points of your unknown.

EXPERIMENT 5

EXTRACTION

Objective

- To introduce an example of the isolation of an organic product from its parent matrix.
- To combine the techniques of extraction, melting points, and acid–base properties you have learned to carry out the isolation.

Background

I. *Review liquid–liquid extractions* discussed in Experiment #1 (Distribution Coefficients).

II. *Extraction of caffeine*

Caffeine is one of the world's most widely consumed food ingredients. It is found in the seeds, nuts, or leaves of plants. The most common source of caffeine is coffee beans, which consist of about 1–4% caffeine. Daily consumption of caffeine has been shown to have some health benefits, as caffeine stimulates respiration, the heart and central nerve system, and promotes urination. However, like many drugs, it can be addictive. Moreover, too much caffeine can cause adverse effects. More details about caffeine's property can be found in a review article (*1*).

Caffeine is consumed frequently in beverages such as coffee, soft drinks, and tea, as well as in products containing cocoa or chocolate. Caffeine concentration varies in different drinks, with coffee having the highest content, which is followed by tea and soft drinks. Table 1 lists the typical caffeine content for several drinks.

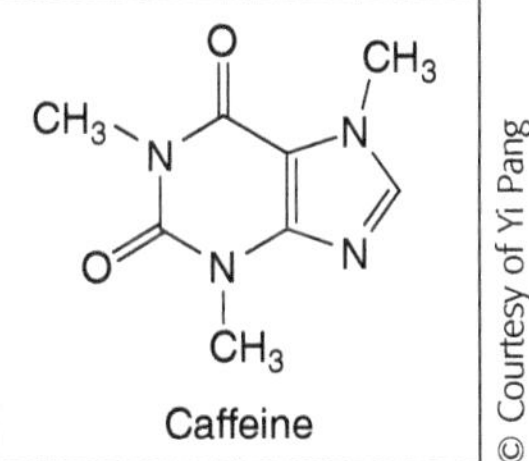

Caffeine

Table 1. Caffeine Content in Common Beverages.

BEVERAGE	CAFFEINE
Coffee, decaffeinated	5 mg (per 8 oz)
Coffee, instant	93 mg (per 8 oz)
Coffee, plain, brewed	133 mg (per 8 oz)
Tea, brewed	53 mg (per 8 oz)
Green tea	45 mg (per 8 oz)
Coca-Cola Classic	23 mg (per 8 oz)
Sunkist	28 mg (per 8 oz)
Diet Coke	31 mg (per 8 oz)
Energy drinks	70–80 mg (per 8 oz)

Data in the table are from reference [1].

The structure of caffeine contains four nitrogen atoms and two carbonyl groups that are polar, and two methyl groups that are nonpolar (hydrophobic). Caffeine exhibits low solubility in water, 2.2 mg/mL at 25 °C. Its water solubility significantly increases with temperature, reaching 670 mg/mL at 100 °C. Caffeine is quite soluble in CH_2Cl_2, making it a convenient solvent for extraction at room temperature. Since caffeine exhibits a large solubility difference in water and CH_2Cl_2, most caffeine will be extracted into CH_2Cl_2 from water at room temperature.

During the extraction experiment in this lab, an aqueous caffeine solution is mixed with another solvent (e.g., CH_2Cl_2) which is *immiscible* with water. The two *immiscible* liquids can be mixed via either shaking or stirring. During mixing, caffeine will move from the aqueous layer to the CH_2Cl_2 layer (until reaching an equilibrium), because it is less soluble in the aqueous layer. Separation of the two immiscible layers on a separatory funnel will allow one to isolate caffeine.

Caffeine is a white crystalline solid that has a melting point of 238 °C. However, caffeine tends to sublime at 90 °C (*2*), making the melting point measurement a challenging task.

III. *Caffeine salicylate salt*

Due to sublimation that occurs below the melting point of caffeine, it is very difficult to take caffeine's melting point in an open capillary tube. One way to overcome this problem is to prepare a caffeine derivative, whose properties can be used to confirm the structure of the original organic compound (i.e., caffeine in this case).

An organic base such as pyridine can react with an acid (such as benzoic acid) to form a salt. In the reaction, the lone pair electron on a nitrogen atom is used to accept a proton.

Pyridine + Benzoic acid → Salt

© Courtesy of Yi Pang

Since caffeine is an organic base, it can also react with an acid to form an organic salt. Because the resulting salt is less soluble in an organic solvent, it will precipitate from methylene chloride. The salt formed from salicylic acid has a sharp melting point and can then be used to characterize caffeine.

Caffeine mp 238 °C + Salicylic acid → Caffeine salicylate mp 137 °C

This nitrogen is more basic

© Courtesy of Yi Pang

References

(1) Heckman, M. A.; Weil, J.; de Mejia, E. G. Caffeine (1, 3, 7-Trimethylxanthine) in Foods: A Comprehensive Review on Consumption, Functionality, Safety, and Regulatory Matters. *J. Food Sci.* **2010,** *75* (3), R77–R87.
(2) CRC Handbook of Chemistry and Physics, 90th ed.; CRC Press: Boca Raton, FL, 2009; pp. 3–84.

IV. *Calculations for theoretical and percent yields*

For a synthetic reaction, it is important to know how much product(s) one obtains from a given reaction. This information is called the "reaction yield".

(a) *Theoretical yield.* The *theoretical yield* of a reaction is the amount of product that would be formed if the reaction proceeds perfectly (an ideal situation). It is the maximum yield one can expect from a reaction. In order to calculate the theoretical yield, the first step is to identify the limiting reagent. This can be demonstrated by considering the following reaction.

$$CH_3CH_2CH_2CH_2OH + NaBr \xrightarrow{H_2SO_4} CH_3CH_2CH_2CH_2Br$$

	$CH_3CH_2CH_2CH_2OH$	NaBr	H_2SO_4	$CH_3CH_2CH_2CH_2Br$
Amount used	8.0 g	13.3 g	20.0g	
Mw	74.12	102.91	98.08	137.03
moles	0.11	0.13	0.20	

In the above reaction, 1-butanol (8 g, 0.11 mol) is used to react with sodium bromide (NaBr) (13.3 g, 0.13 mol) to form 1-bromobutane. In the reaction, 1-butanol limits the extent of the reaction, as NaBr is in excess.

Limiting reagent is the chemical (or reactant) that determines the extent of the reaction (or how far the reaction can go). The first step in calculating theoretical yield is to identify the "limiting reagent." In the above experiment, 1-butanol (0.11 mol) is the limiting reagent, since the reaction uses a slight excess of sodium bromide (0.13 mol, about 1.18 equiv).

Thus, the theoretical yield of product (i.e., 1-bromobutane) is 0.11 mol. The maximum weight of 1-bromobutane is then calculated as shown below:

0.11 (mol of butanol) × 137.03 (MW of product)
= 14.8 g (1-bromobutane) (theoretical yield of product)

(b) *Percent yield.* An organic reaction usually does not have 100% yield, unless the reaction gives only one pure product. Most organic reactions produce at least one by-product in addition to the desired product. An important task in performing an organic experiment is to obtain the highest yield for the desired product. If the actual yield of 1-bromobutane is only 10.5 g, the percent yield can be calculated as shown below.

$$\text{Percent yield} = \frac{\text{Actual yield (in gram)}}{\text{Theoretical yield (in gram)}}$$

$$= \frac{10.5\text{ g}}{14.8\text{ g}} \times 100 = 71\%$$

The reaction is then said to proceed with a 71% yield.

Experiment

- Obtain 10 mL cola syrup from Stockroom with a 50 mL beaker.
- Into a clean 125 mL separatory funnel, pour the 10 mL of cola syrup and rinse the beaker with 10 mL of water.
- Add 2 mL of ammonium hydroxide and swirl to mix the liquids.
- Add 10 mL of methylene chloride, stopper the separatory funnel and shake *gently* to avoid the formation of an emulsion, and then release the pressure. Repeat this over a period of 10 min.

- Allow the separatory funnel to stand until the two layers have separated (this may take several minutes).
- Carefully draw off the bottom layer into a small Erlenmeyer flask (do not allow water to enter the stop-cock).
- Extract the aqueous mixture with another 2 mL of methylene chloride, as above, and transfer the organic layer to the Erlenmeyer flask with the first extract.
- Add 0.5 g [**ESTIMATE**, Never weigh drying agents on the electronic balance] of anhydrous sodium sulfate to the organic solution in the Erlenmeyer flask, and swirl for a few minutes (density of anhydrous Na_2SO_4 is 2.664 g/cm^3. 0.5 g of anhydrous Na_2SO_4 is about 0.19 mL).
- When the solution is clear and there are no droplets of water visible, transfer the liquid with a Pasteur pipette to a preweighed 25 mL round-bottom flask (make sure none of the drying agent gets into the flask).
- Place the round-bottom flask on a *warm* steam bath and evaporate the methylene chloride.
- Wipe all traces of moisture from the flask and weigh the flask to determine the weight of caffeine (there should be 5–18 mg).
- If you have a different weight of caffeine, see your TA.
- The caffeine is not sublimed under the experiment conditions we used.

Caffeine Salicylate

Caffeine (mp 238 °C) + Salicylic acid → Caffeine salicylate (mp 137 °C)

- Weigh out the amount of salicylic acid equivalent to the number of moles of your sample of caffeine.
- Dissolve the caffeine in 1 mL of methylene chloride.
- Add the salicylic acid and swirl to dissolve.
- Let stand for 1 min.
- Add hexane dropwise until the solution just begins to appear cloudy when a drop of hexane is added.
- Chill the mixture in an ice-water bath.
- When crystallization is complete, carefully remove the liquid using a disposable pipette.
- Wash the solid with a little cold hexane.
- Allow the solid to stand in air until it is dry (overnight if possible) or vacuum dry the solid.
- Determine the weight and turn in the sample.
- Obtain a melting point of the product.

Laboratory Notebook

- Use the format outlined in the syllabus.
- Full prelab.

Product

- Caffeine salicylate that you isolate and its melting point.
- Calculations for theoretical and percent yields.

Experiment 6

FRACTIONAL DISTILLATION

Objective

- To introduce the use of distillation to purify an organic liquid.
- To show the advantages of a fractional distillation over a simple distillation.

Background

The first distillation most likely happened when an ancient chemist discovered that heating an alcohol-containing solution and condensing the vapors into a different container than the one being heated resulted in a more potent brew. Ethanol (bp 78.1 °C) boils at a lower temperature than water (bp 100 °C), and so this early distillation resulted in the condensate having a higher concentration of ethanol than the liquid from which it was being heated/distilled. Interestingly, the temperature that a water–ethanol solution distills is ~78.3 °C, and the distillate is only ~95% ethanol (and 5% water), not 100% ethanol.

In Experiment 3, we discussed how intermolecular forces (forces between two or more molecules) determine the physical properties of a given molecule. For a liquid, these forces define at what temperature the liquid undergoes the transition to a gas. Boiling, and as a result distillation, occurs when these intermolecular forces are disrupted upon the application of energy (heat). Disruption of the forces results in some of the molecules at the liquid–gas interface (i.e., the surface of the liquid) transitioning into a vapor above the liquid (the so-called vapor phase). This equilibrium between the liquid solution and the gaseous vapor above the liquid can be measured and is called the vapor pressure. Increasing the number of molecules in the vapor phase increases the vapor pressure, and decreasing the amount decreases the vapor pressure.

Simple and Fractional Distillations

We will use two different types of distillation: (1) a simple distillation and (2) a fractional distillation. A simple distillation (Figure 1, left) uses a distillation flask, a thermometer to measure distillation temperature, a condenser, and a receiving flask/container. Heating the distillation flask boils the liquid inside the flask, and the vapors either condense back into the distillation flask or are directed into a condenser. The liquid is cooled in the condenser and flows into a receiving container. The simple distillation setup works well for separating two liquids having a boiling point difference of greater than 75 °C. Distillation of two liquids having boiling points closer than that usually results in a mixture of the two components.

A fractional distillation (Figure 1, right) uses an additional column/tube between the distillation flask and the condenser and does a much better job of separating two liquids having boiling points closer together than 75 °C. Fractional distillations are generally characterized by the column separating the distillation flask from the condenser. A distillation using the setup shown in Figure 1 (right) proceeds when the vapor generated on heating "climbs" the column. This climbing process is incremental, with the initial vapors condensing on the cooler glass and reentering the liquid phase in the distillation flask. As the column progressively heats up, the vapor becomes enriched with the lower boiling liquid, which selectively climbs higher up the column and eventually begins to transition into the condenser. The climbing process proceeds by a series of condensation–vaporization steps, each of which separates the two different liquids slightly and is called a *theoretical plate*. While the column does not have to contain anything (as shown in Fig. 1, right), most of the time this column is packed with something that is inert but that also increases the surface area inside the column. The increased surface area results in a more efficient distillation by increasing the number of vaporization–condensation cycles (i.e., it increases the number of theoretical plates) of the material that helps to slow down the vapor as it climbs toward the condenser. Common packing materials include glass beads or helices, metal sponges, or other material. It is very important when performing a fractional distillation that the rate of distillation is kept low, so that thermal equilibrium can be attained at each theoretical plate.

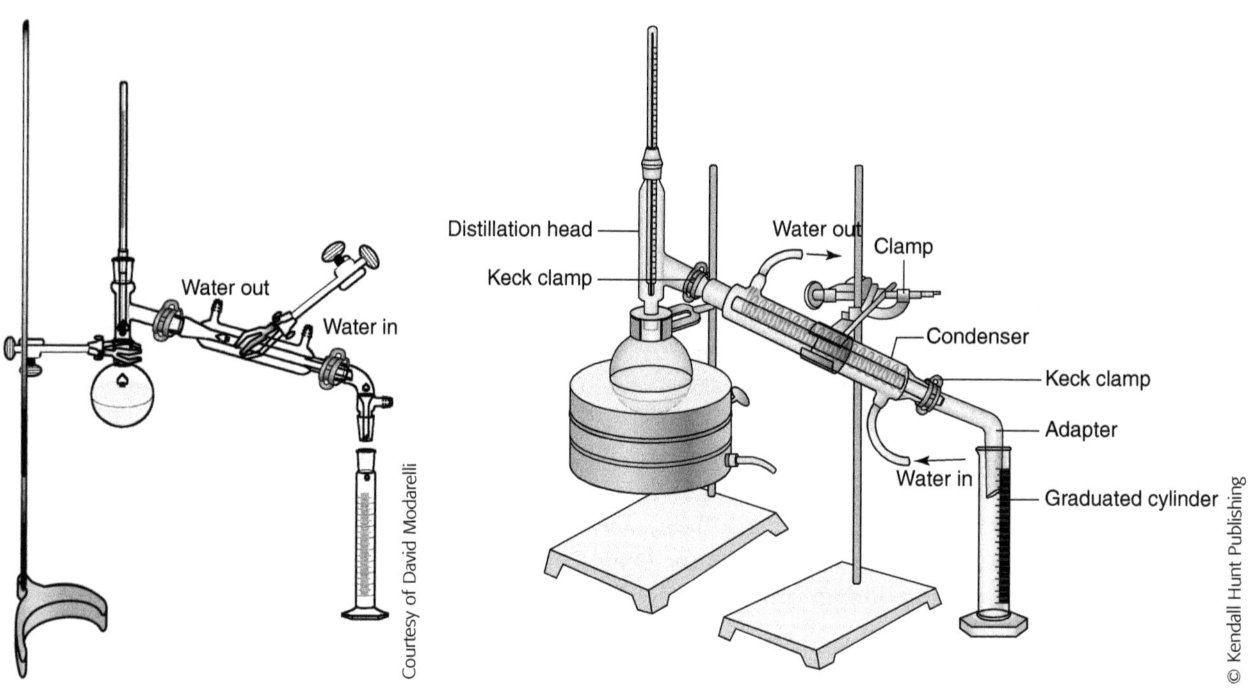

Figure 1. Simple and fractional distillation systems.

Boiling Points and Distillation

We will normally view a change in boiling point during a distillation as the point when we stop collecting the first component of a mixture and start collecting the second component. This scenario will not always be the case. Two examples are that of *azeotropes* and the presence of two similarly boiling components. An azeotrope is a mixture of two liquids that co-distill together in a certain component mixture. An example of such a solution is a mixture of ethanol and water. As a result of hydrogen-bonding interactions, this mixture boils at a constant temperature of 78.1 °C and is composed of 95.5% ethanol (bp 78.3 °C) and 4.5% water (bp 100.0 °C). Even with a mixture that is, for example, 80% water and 20% ethanol, the distillation will occur at 78.1 °C in a 95.5/4.5 ratio until the ethanol is exhausted, at which point water will be the only component and will then boil at 100 °C.

The boiling point of a liquid is also dependent on the ambient pressure, since the temperature and pressure of an ideal gas (and therefore the vapor pressure) are related to one another by the ideal gas law. The ambient pressure is therefore generally reported when a boiling point is reported. The ideal gas law can be used to decrease the distillation temperature of a liquid. This effect is usually accomplished by performing the distillation at a reduced pressure, decreasing the boiling point of the liquid and potentially minimizing heat-induced sample degradation.

Experiment

- Check out a 100 mL round-bottom (boiling) flask, a 50 mL graduated cylinder and a second condenser from the stockroom.
- Construct the "simple" distillation apparatus shown in Figure 1 (left). You will be using the graduated cylinder as a receiver in this distillation. Be sure to apply *a small amount* of vacuum grease to ground glass joint connections to prevent them from seizing. You should also use the metal clamps to connect your horizontal condenser and the round bottom flask to the monkey bars in your hood. When your distillation apparatus is assembled, check for any gaps at the ground glass joint connections to make sure there are no leaks in the setup.
- You will be distilling 50 mL of the premade 50:50 mixture of water and acetone. You will record the temperature of distillation, as indicated by the thermometer, for every 2 mL of distillate collected, and produce a plot similar to the curve (solid line) shown in Figure 2 for this simple distillation. The controller for the heating mantle should be set on HIGH.
- Collect ~28 mL of distillate.
- Allow the apparatus to cool. Return the distillate to the distillation flask and modify the apparatus to perform a fractional distillation using the packed condenser obtained from the stockroom (Figure 1 right).
- Repeat the distillation and record the temperature of distillation for every 2 mL of distillate collected and produce a second plot similar to the curve (broken line) shown in Figure 2 for this fractional distillation.
- Collect ~26 mL of distillate.

  Note: *When you finish with your final distillation, place your acetone (after measuring the amount in your graduated cylinder) into the designated container.*

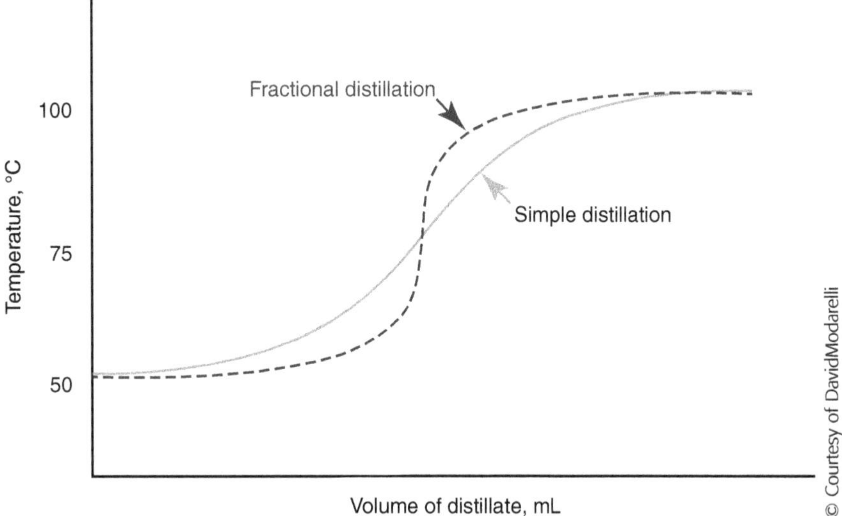

Figure 2. Distillation curves for an acetone–water mixture using a simple (solid line) and fractional distillation system (dotted line). In the fractional distillation, the separation between the two different boiling solvents is more well defined, and the separation is therefore better.

Laboratory

- Use the format outlined in the syllabus.

Laboratory Notebook

- There are no reactions or mechanisms for this experiment.

Product

- On separate paper attached to your notebook: The two plots of temperature versus distillate collected for the simple and fractional distillations.

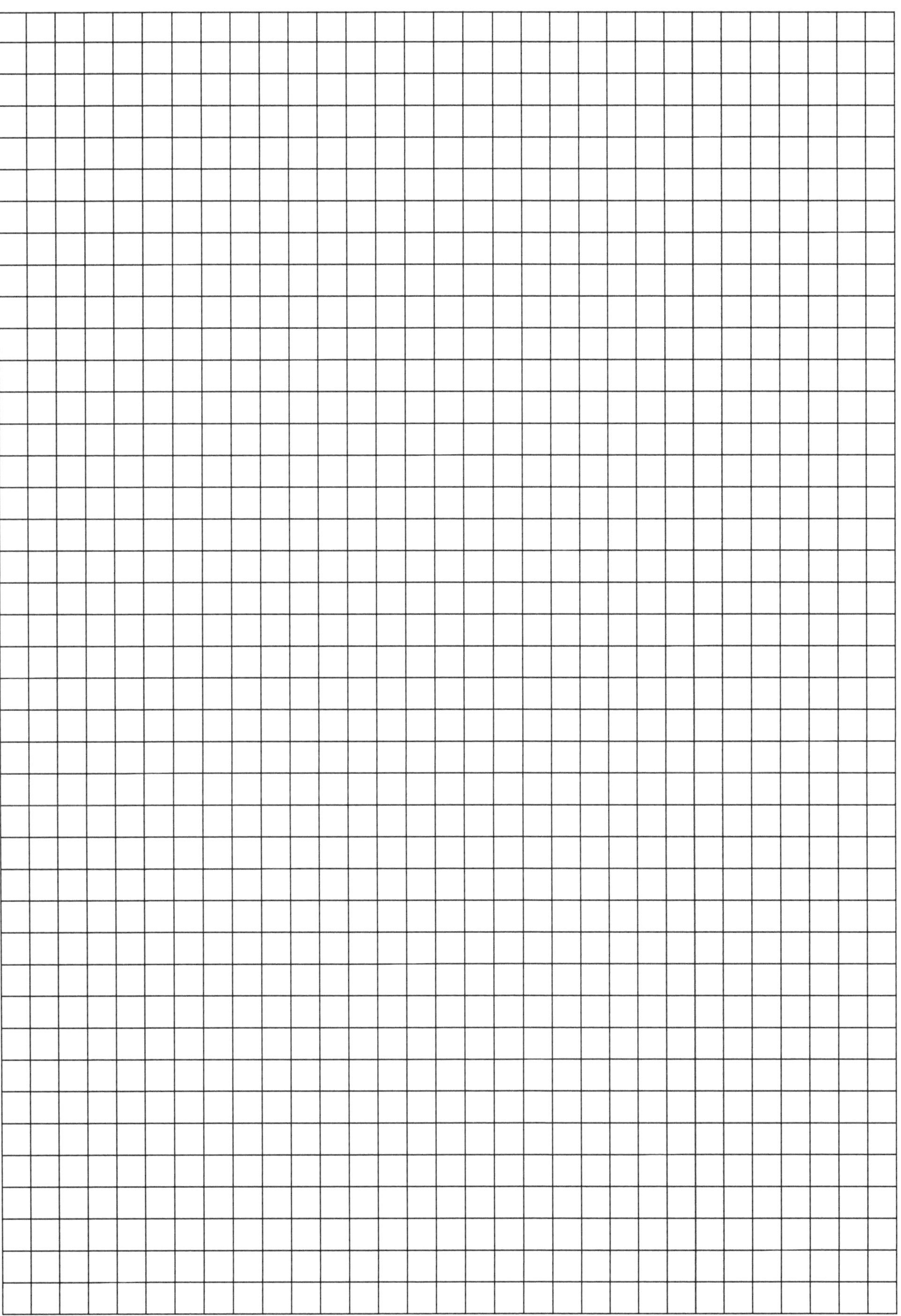

EXPERIMENT 7

METHYLCYCLOHEXENES

Objective

- To introduce a functional group transformation procedure, utilizing the skills you have learned.

Background

The acid-catalyzed dehydration of alcohols is an important reaction in organic chemistry because it facilitates the formation of C=C double bonds. The acid of choice is often phosphoric acid. Sulfuric acid can also be used but is sometimes problematic because it can cause polymerization and the formation of other byproducts, which can result in a depressed yield of the desired olefin.

This reaction is typically an E1 elimination. Tertiary alcohols are easier to dehydrate than secondary alcohols, followed by primary alcohols (usually requires a higher temperature to proceed and proceed by an E2 mechanism). The reaction scheme is generally considered as a three-step mechanism:

- Protonation of the hydroxyl group:

H OH $\underset{}{\overset{H_3O^+}{\rightleftharpoons}}$ H $\oplus OH_2$

- The departure of water to form the carbocation intermediate:

H $\oplus OH_2$ $\rightleftharpoons$ H $\oplus$ + H_2O

- Deprotonation on an adjacent carbon to furnish the olefin products:

H_3O^+ + H (minor product) $\rightleftharpoons$ H–Ö: (H) H $\oplus$ ⟵ H $\oplus$ ⟶ H–Ö: (H) H $\oplus$ $\rightleftharpoons$ (major product) + H_3O^+

minor product

major product

In scenarios where different regioisomers can form (as demonstrated above), the major product will be the more substituted alkene as predicted by Zaitsev's rule, which states that the more substituted alkene is the more thermodynamically stable alkene.

In this experiment, we will explore the dehydration of 2-methylcyclohexanol. The dehydration reaction will be carried out under acidic conditions to yield a mixture of two isomers of 2-methylcyclohexene (Figure 1). By applying Zaitsev's Rule, we can predict the major and minor products from this reaction.

Figure 1. The dehydration of 2-methylcyclohexanol.

Experiment

- Into a clean, preweighed 25 mL round-bottom flask get 5.0 mL of 2-methylcyclohexanol from the stockroom. Reweigh and obtain the weight of the compound.
- There is significant glassware breakage in this experiment. Thermometer, distillation head, and receiving flask.
- To the flask, add 1.5 mL of phosphoric acid and a boiling chip.
- Assemble the distillation apparatus as shown in the illustration below.

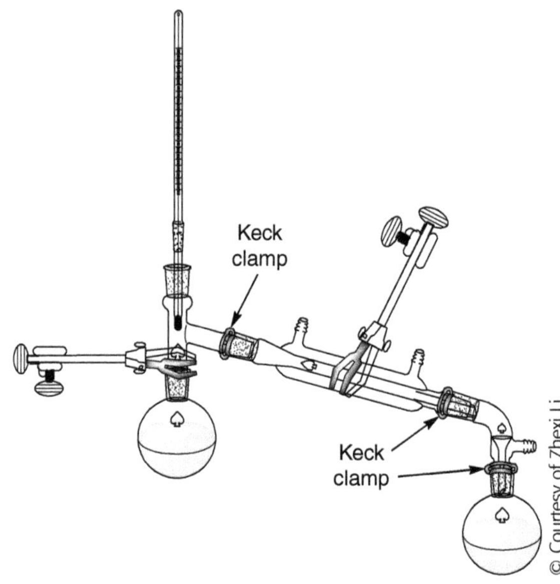

- Be sure your clamps are connected to the monkey bars in your hood.
- Heat the flask with a heating mantle. Set the controller of the heating mantle to HIGH.
- ***Do not allow the temperature of the vapor to rise above 130 °C.***
- Distill until only a *small amount of liquid remains.*
- Allow the apparatus to cool.
- The distillation receiver contains the alkenes and water.
- With a Pasteur pipette draw off as much of the aqueous layer as you can.
- The distillation residue for the first step should be neutralized with NaOH (calculated amount based on the phosphoric acid). If the organic layer is visible, separate the organic layer from the aqueous. The aqueous layer goes to sink. For the second step, the distillation residue should go into the organic non-halogenated waste.

- Add two drops of 10% sodium carbonate solution to the organic layer (caution—CO_2 may be evolved) until the aqueous layer is basic to pH paper.
- Again draw off the aqueous layer with a pipette.
- To the organic layer add 0.5 g of solid calcium chloride [ESTIMATE—*Do Not weigh drying agents on the electronic balance!*], and swirl occasionally until the liquid is dry and clear.
- If the liquid is not clear within 5 min, add another 0.2–0.3 g of $CaC1_2$.
- With a Pasteur pipette, transfer the alkenes into a clean 10 mL flask.
- If the alkenes must be stored overnight before distillation, stopper the flask with the glass stopper, lightly greased with stopcock grease.
- Before distilling the product, preweigh the 10 mL receiving flask and clean the distillation apparatus.
- Distill using short path distillation:

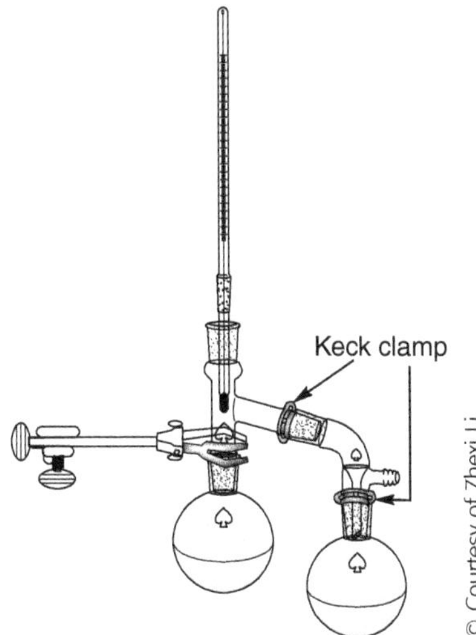

- Determine the weight of the product by weighing the receiver flask with the contents.
- SHOW your TA your isolated product in the receiving flask BEFORE emptying it out. If you do not do so, you must repeat the experiment.

Laboratory Notebook

- As per the syllabus.
- Full prelab.
- Make sure you have the boiling points and densities for all the materials used in this experiment, 2-methylcyclohexanol and the two isomers of 2-methylcyclohexene produced.

Product

- Your methylcyclohexene product.
- Calculations of limiting reactant, theoretical yield, and percent yield for the reaction.

Experiment 8

CATALYTIC HYDROGENATION

Objective

- To demonstrate the use of a catalyst in the organic laboratory.
- To demonstrate the use of a compressed gas as a reactant.
- To demonstrate the use of the Brown apparatus.

Background

Catalytic hydrogenation is a hydrogenation reaction carried out in the presence of a catalyst such as nickel, palladium, or platinum. The reaction is usually carried out by using H_2 gas. Addition of hydrogen to alkenes is one of the most common reactions, and is widely used in both laboratory and industry settings. A significant advantage of using a gaseous reactant is that the product is easy to purify.

An alkene + H_2 $\xrightarrow{\text{Catalyst}}$ An alkane

In order to maximize the surface area for hydrogen and alkene adsorption, the metal catalyst is usually supported on a porous inert material such as carbon or charcoal. An example is palladium on carbon, Pd/C, which is commonly used in laboratory. During the reaction, the alkene and hydrogen are absorbed onto the catalyst surface, which is followed by the hydrogen transfer (from the catalyst to the alkene). Hydrogenation usually occurs with *syn* stereochemistry, with both hydrogens adding to the double bond from the same face, as shown in the hydrogenation of 1,2-dimethylcyclohexene below.

$$\text{1,2-dimethylcyclohexene} \xrightarrow[\text{Pd/C}]{H_2} \textit{cis}\text{-1,2-dimethylcyclohexane}$$

The largest scale application of hydrogenation is used in the processing of vegetable oils in the food industry. Vegetable oil typically consists of esters composed of propane-1,2,3-triol, also called glycerol, and three long chain fatty (carboxylic) acids. Olive oil is primarily composed of glycerol trioleate, where the fatty acid components are a mixture of oleic acid (monounsaturated, i.e., one unsaturated C=C bond shown in the structures below), linoleic acid (polyunsaturated) and palmitic acid (saturated) that depends on the region the olive oil was grown in. The oleic acid portions contain a carbon–carbon double bond with the *cis*-configuration. The oleic acid portion does not pack well in a crystal lattice, and olive oil is therefore liquid at room temperature.

Glycerol trioleate (mp −5.5 °C) — H_2, Ni → Partial hydrogenation — H_2, Pd → Glycerol tristearate mp 69.9 °C

Chemical structure of fatty acids
Stearic acid:
Oleic acid:

© Courtesy of Yi Pang

When all carbon–carbon double bonds are reduced to single bonds, the final product is glycerol tristearate. Fully reduced glycerol tristearate is a solid as a result of the reduction of all of the *cis*-CH=CH bonds. Oils are often partially hydrogenated by controlling the extent of the reaction. Since only part of *cis*-CH=CH bonds are reduced, the product contains a mixture of liquid and solid fats, or semisolid fats, which are preferred for baking. In practice, the manufacture of margarine is accomplished by partial hydrogenation of a vegetable oil (e.g., corn or soybean oil), and the product is widely used to produce baked goods. During the partial hydrogenation process, some of the *cis*-CH=CH bonds are also isomerized to the more stable *trans*-CH=CH bonds. Since palladium catalysts have a higher tendency to cause the *cis* to *trans* isomerization, nickel or platinum catalysts can be used instead to reduce the isomerization. Because *trans* isomers contribute to the formation of the so-called "bad fat" LDL, many countries have introduced mandatory labeling of *trans* fats on food products and have appealed to the food industry to institute voluntary reductions.

In a laboratory, conversion of an alkene into an alkane is readily achieved by stirring it under hydrogen gas at room temperature in the presence of a platinum or palladium catalyst. A convenient setup for hydrogenation was initially reported by Brown (H. C. Brown and C. A. Brown, *J. Am. Chem. Soc.* **1962**, *84*, 1495). Since then, it has been widely used in organic laboratories. In this experiment, we will use Brown's apparatus to achieve hydrogenation of an alkene.

Chemical Safety Information. Hydrogen is a colorless and odorless gas, which is flammable and explosive when mixed with air or oxygen.

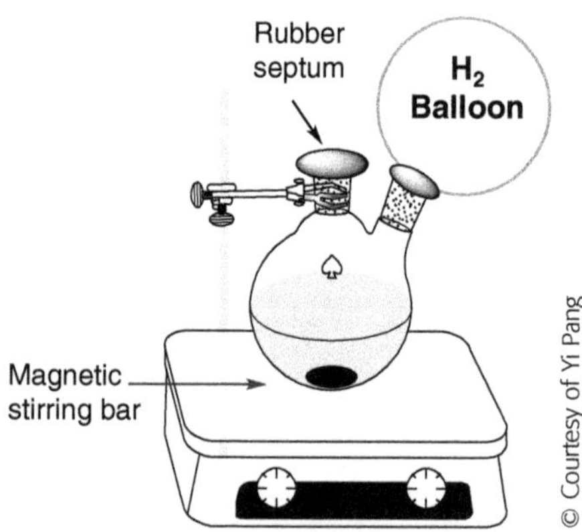

Figure 1. Diagram of Brown's apparatus setup for hydrogenation.

Experiment

You will be converting norbornenedicarboxylic acid to norbornanedicarboxylic acid using catalytic hydrogenation. You will use a magnetic stir plate, you will need to check out a magnetic stir bar, balloon, rubber band, and septum from the stockroom.

CO_2H CO_2H —— H_2, 10% Pd/C / $CH_3CO_2CH_2CH_3$ ——→ CO_2H CO_2H

© Courtesy of Yi Pang

- Set up the Brown apparatus as shown in discussion. Also, see the set-up apparatus outside the stockroom. Be sure to have the balloon secured using a rubber band (otherwise, the apparatus will leak).
- Weigh out 90–100 mg of norbornenedicarboxylic acid and load this into a 25 mL round-bottom flask. (Record the exact weight.)
- To this flask, add about 10± mg of the 10% Pd/C catalyst. (Do not need weigh. There is a sample tube containing the estimate amount)
- To this flask, carefully add 5 mL of ethyl acetate and the magnetic stir bar.
- Attach the Brown apparatus to the flask.
- Take the entire setup to the hydrogen tank
- Your TA will then operate the H_2 cylinder and fill the apparatus with H_2 gas. Carry the apparatus back to your bench-top and clamp everything in place as instructed.
- Magnetically stir the contents of the flask for 60 min. This step is to be done in the hood.
- After stirring for 60 min, carefully remove the septum from the apparatus to release excess H_2.
- *Save the 25 mL flask and its contents!*
- Assemble the apparatus for vacuum filtration with the Hirsch funnel (see Figure 2 on page 26). With the vacuum on to provide suction (from pump or aspirator), secure the filter paper over the holes of the Hirsch funnel. Then add to the funnel about 0.04 g of filter aid and rinse this with 1 mL of ethyl acetate (you only need a thin layer of filter aid to cover the surface of the filter paper. Do not need to weigh).
- Vacuum filter the contents of the 25 mL flask. Wash the flask with 1 mL of ethyl acetate and vacuum filter the rinse. Your product is the filtrate (collected liquid) and should not contain any black particles, coloring from the catalyst, or filter aid. If so, refilter with fresh filer aid. (Please secure the filtration flask as instructed, as many Hirsh funnels are broken during this step in the past, which results in loss of your product).
- Evaporate the filtrate in a preweighed 25 mL round-bottom under reduced pressure.
- Obtain the weight and melting point of the resulting solid.
- Return all borrowed materials to the stockroom.

Use modified set-up in the experiment

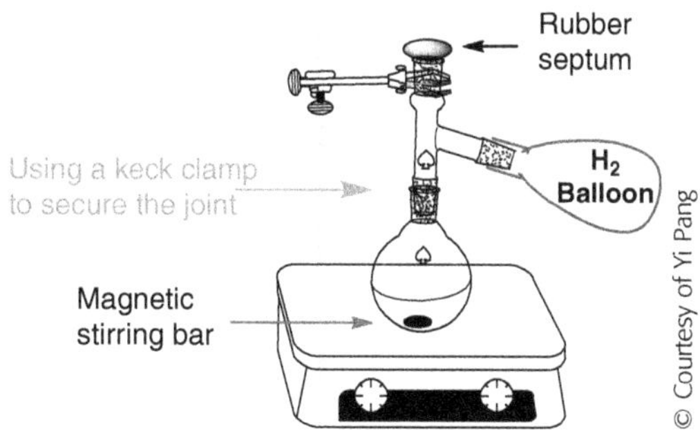

© Courtesy of Yi Pang

Laboratory Notebook

- As per the syllabus.
- Full prelab, no mechanism.
- Remember to have melting points and molecular weights for norbornenedicarboxylic acid and norbornanedicarboxylic acid.

Product

- Calculations for theoretical and percent yields. Be sure to indicate the limiting reagent.
- Your norbornanedicarboxylic product and its melting point.

Experiment 9

SYNTHESIS OF 1-BROMOBUTANE

Objective

- To learn how to use a reflux apparatus and to perform a S_N2 reaction in an organic synthetic reaction.

Background

Please see the section on S_N2 reactions in your textbook for background information regarding S_N2 (and S_N1) reactions.

This reaction will involve the conversion of 1-butanol to 1-bromobutane using an S_N2 reaction. We will use a combination of sodium bromide and sulfuric acid to perform this reaction. Primary alcohols will often undergo this S_N2 reaction well. Under the strongly acidic conditions used here, secondary and tertiary alcohols will generally form a carbocation, and either undergo a substitution reaction via an S_N1 mechanism or undergo an elimination reaction to make an alkene (via an E1 mechanism).

$$\text{OH} \xrightarrow[H_2SO_4,\ \Delta]{NaBr} \text{Br} + H_2O$$

Many organic reactions do not occur quickly at room temperature and are therefore heated to increase the rate at which product formation occurs. Because temperature and pressure are related, heating a reaction solution in a closed container will also increase the pressure inside that container, resulting in an explosion. However, heating an open container will result in evaporation of that solvent. In order to avoid these problems, we will *reflux* our solution. Heating a sample at reflux (i.e., at the boiling temperature of the solvent) using the apparatus shown in Figure 1 allows us to heat the sample, while avoiding solvent loss. The cold water flowing through the input and output tubes (i.e., the water jacket of the condenser) cools the heated solvent vapors, condensing them back into the reaction flask and preventing the solvent from boiling away out of the top of the condenser. We will use a slow flow of water through the condenser to avoid wasting water *and* popping the water tubes off our condenser. The amount of heat we will apply will be enough so that the solvent vapor rises to bottom third of the condenser, but not higher.

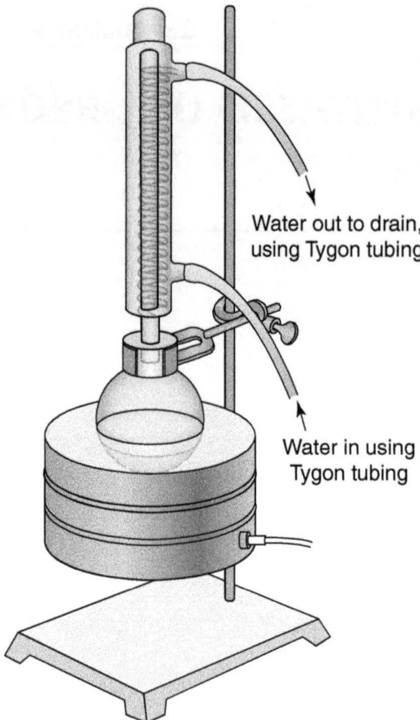

Figure 1. Apparatus used for heating a reaction mixture to reflux (boiling).

Experimental Procedure

- Add 2.4 g of 1-butanol into a 25 mL round-bottom flask. You can add 1.48 mL twice by using a syringe pipette (do not use balance).
- Add 4.0 g of sodium bromide, 4.5 mL of water, and 3.3 mL of concn H_2SO_4 to the 25 mL round-bottom flask. Add one or two Teflon boiling chips. Set up the apparatus for reflux as shown in Figure 1. The concentrated H_2SO_4 can be distributed by TA with a pipette syringe.
- With a gentle flow of water through the condenser, heat the reaction mixture to reflux with a heating mantle for a period of 60 min. The heating mantle controller can be set to ~4. Remove the heat and allow the apparatus to cool. Observe what changes have occurred to the reaction mixture.
- Replace the reflux condenser with the short-path distillation apparatus (the same one you used for the methylcyclohexenes experiment, Experiment 7). Set the heating mantle controller to HIGH. Cool the receiving flask in a beaker of ice water.
- Distill until only water comes over. The temperature should not exceed 100 °C.
- The distillate will contain 1-bromobutane, water, and several byproducts of the reaction.
- Remove the lower (1-bromobutane) layer with a disposable pipet and transfer it to a dry Erlenmeyer flask.
- Add ~0.5 g of anhydrous sodium sulfate (estimate, do not weigh out), mix well, and let set for several minutes. If the appearance of the sodium sulfate is wet, use an additional amount of sodium sulfate to complete the drying.
- Transfer (using a clean pipette) the 1-bromobutane to a clean, dry 10 mL boiling flask, and set up the apparatus for another short path distillation, making sure you preweigh the receiving flask.
- Distill your product, and show it to your TA when the distillation is complete.

Laboratory Notebook

- Use the format outlined in the syllabus.
- Full prelab.

Product

- 1-Bromobutane product
- Calculations of limiting reactant, theoretical yield, and percent yield for the reaction

Experiment 10

DIPHENYLACETYLENE

Objective

- To illustrate the reactivity of alkenes and their conversion to alkynes.

Background

In this experiment, you will synthesize diphenylacetylene from *E*-stilbene in two steps. *E*-Stilbene will first be brominated to form *meso*-stilbene dibromide, followed by dehydrohalogenation to furnish the desired alkyne.

Halogenation of Alkenes

Olefins easily undergo addition reactions when treated with halogens (such as bromine). In this experiment, *E*-stilbene is treated with pyridinium hydrobromide perbromide instead of liquid bromine because liquid bromine is difficult to handle and extremely toxic. Pyridinium hydrobromide perbromide is an odorless, crystalline solid that generates bromine in situ, making it a safe alternative to liquid bromine for this experiment. The general reaction scheme for the bromination is illustrated below:

- Formation of the bromonium intermediate:

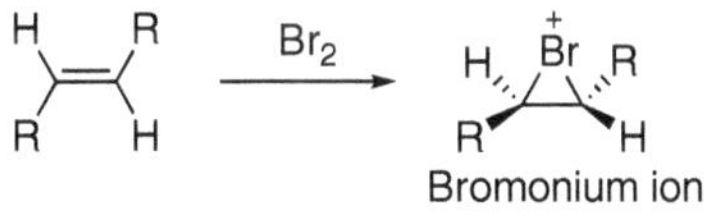

Br_2

Bromonium ion

- Attack of bromide on the bromonium to furnish the vicinal dibromides (a *meso* compound):

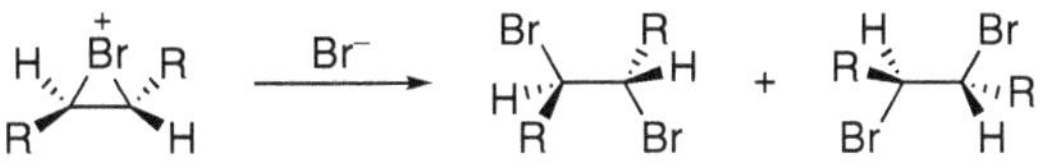

Br^-

A characteristic feature of this bromination is that the bromides will add to the opposite sides of the carbon–carbon double bond. This addition is called *anti* addition. Because the *anti* adduct is preferentially formed over the *syn* adduct, this reaction is said to be *stereoselective*.

Dehydrohalogenation of Vicinal Dihalides

The second step of this experiment involves the dehydrobromination of the *meso*-stilbene dibromide to afford the desired diphenylacetylene:

$$\text{Ph(Br)CH–CH(Br)Ph} \xrightarrow[\text{Triethylene glycol}]{\text{KOH}} \text{Ph–C}\equiv\text{C–Ph}$$

KOH acts as the base to remove the acidic protons from *meso*-stilbene dibromide, and this reaction proceeds through an E2 mechanism. Because of the high temperature required for the dehydrohalogenation to proceed, triethylene glycol is used as the solvent (bp 285 °C).

Experiment

meso-Stilbene Dibromide

- In a reaction tube, dissolve 60 mg of *E*-stilbene in 1.2 mL of acetic acid by heating on a steam bath or a hot water bath.
- Add 120 mg of pyridinium perbromide to your reaction tube. [There will be an example sample of 120 mg of pyridinium perbromide in the chemical hood. Use this to *estimate* 120 mg for your reaction (**DO NOT** weigh this reagent)].
- Mix the pyridinium perbromide in by swirling. If necessary, rinse crystals of the reagent down the wall with a little acetic acid. Heating the mixture on a steam bath or hot water bath (for about 2 min), cool the mixture.
- Procedure for the isolation of *meso*-stilbene dibromide: Cool the reaction mixture under tapwater. Use a pipette to separate and place the solution in an Erlenmeyer flask (this solution may contain excess pyridinium perbromide). Wash the product with methanol (~1 mL) and collect the product on a Hirsch funnel. Use a pipette to remove all the solvent from the vacuum flask and place this solvent in the flask labeled "pyridinium perbromide waste" at the side of the lab. A small amount of additional methanol can be used to wash and remove any color. The yield of colorless crystals (mp 236–237 °C) should be ~80 mg.
- Take a small sample (5 mg) of the dibromide for a melting point and to submit as the product. Use the remaining product for the next step.
- Since the melting point is above 200 °C, use a Melt-Temp apparatus for determining the melting point of your sample. (Note: Remember to turn the power to low after using the Melt-Temp melting point apparatus.)

Clean Up. To your flasks that contained the pyridinium perbromide, add sodium bisulfite (until a negative test with starch–iodide paper is observed and the orange color disappears) to destroy any remaining perbromide. Neutralize this solution with sodium carbonate and extract the pyridine released with ether. This ether layer goes in the organic solvent waste container.

Diphenylacetylene

- Place all the remaining dibromide (weigh it) in a reaction tube.
- To the reaction tube, add 1 mL of triethylene glycol and one pelletof KOH, regardless of the weight of the dibromide.
- Heat the mixture in a sand bath. Try to make sure that the solution becomes homogeneous at ~100°C, before taking it up to 160–170 °C. Keep the reaction at 160–170 °C for 5 min. You may have to take the test tube out of the sand bath and put it back in repeatedly.

- Cool the mixture to room temperature, remove the thermometer, and add 2 mL of water. Make sure to stir the mixture well after adding water to the cooled reaction mixture.
- Collect the diphenylacetylene product, which is a colorless granular solid, using a Hirsch funnel.
- Recrystallize your product in a preweighed 10 mL round bottom flask. (You may use 95% ethanol.)
- Obtain the melting point of the product.

Laboratory Notebook

- Use the format provided in the syllabus packet.
- Full prelab.

Product

- The properly labeled solid products and their melting point.
- Calculations of limiting reactant, theoretical yield, and percent yield for both reactions.
- Equation and mechanism for both reactions.

EXPERIMENT 11

SELECTIVE REDUCTIONS

Objective

- To illustrate how an organic chemist can selectively react a molecule with more than one functional group.

Background

Very often, an organic molecule contains two or more functional groups. By choosing the appropriate reagent and reaction condition, one can achieve selective reaction on a specific functional group in the molecule. The following reaction sequence illustrates an industrial process that is used to produce ibuprofen, a nonsteroidal, anti-inflammatory drug (NSAID) developed in late 1960s. The first reaction in the process is hydrogenation in the presence of a catalyst. In the reactant (4-isobutylphenyl) methyl ketone, there are two functional groups (i.e., phenyl and carbonyl) that can be reduced. By using Raney nickel as the catalyst, the catalytic hydrogenation only occurs on the carbonyl group without reducing the benzene ring.

(4-isobutylphenyl) methyl ketone $\xrightarrow[\text{Raney nickel}]{H_2}$ (alcohol) $\xrightarrow[\text{"Pd" catalyst}]{CO}$ Ibuprofen

Selective reduction

Because most molecules contain many functional groups, the success or failure of a lengthy organic synthesis scheme depends on the ability of a chemist to choose the appropriate reagents and conditions to transform only the functional group necessary, without affecting the other groups in the molecule.

In a laboratory setting, selective reduction of a carbonyl group can be conveniently achieved by using sodium borohydride, $NaBH_4$. At room temperature in ethanol, $NaBH_4$ rapidly reduces aldehydes and ketones without reducing other functional groups (e.g., epoxides, esters, and nitro groups). Since $NaBH_4$ only reacts slowly with alcohols, the reduction with $NaBH_4$ is often carried out in ethanol.

Sodium borohydride

In this experiment, you will actually perform two "sub-experiments" on the same molecule, *m*-nitroacetophenone, which contains two functional groups that can be reduced.

In the first experiment, you will reduce the ketone using sodium borohydride in ethanol, without reducing the nitro group contained in the molecule. In this reaction, the hydride (a hydrogen with two electrons and a negative charge) is transferred from the borohydride to the carbonyl carbon, followed by protonation in the work-up after the reaction to form an alcohol. This effectively adds the elements of H_2 across the carbon–oxygen double bond, similar to the catalytic hydrogenation of norbornenedicarboxylic acid you did in Experiment 8. The product from this reduction will be an alcohol. Sodium borohydride is relatively expensive, but one equivalent of sodium borohydride can reduce *four* equivalents of a ketone or aldehyde, making it a very economical reagent.

$$4\ (m\text{-}O_2NC_6H_4COCH_3) + BH_4^- \xrightarrow{CH_3CH_2OH} 4\ (m\text{-}O_2NC_6H_4CH(OH)CH_3)$$

In the second experiment, you will reduce the nitro group by utilizing elemental tin and hydrochloric acid, without reducing the ketone that is present in the molecule. Tin has an electron configuration $[Kr]4d^{10}5s^25p^2$, which can lose four electrons to give Sn^{4+} cation. In the following reaction, a total of 12 electrons are transferred from tin (being oxidized—losing electrons) to the nitro group (being reduced—gaining electrons).

$$2\ (m\text{-}O_2NC_6H_4COCH_3) + 3\ Sn + 14\ H^+ \longrightarrow 2\ (m\text{-}H_3N^+C_6H_4COCH_3) + 3\ Sn^{+4} + 4\ H_2O$$

The use of "tin and hydrochloric acid" is a classical method for the reduction of a nitro group to an amine. The method is especially useful for the reduction of an aromatic nitro compound to the corresponding amine, since metal hydrides (such as $NaBH_4$ and $LiAlH_4$) are less effective for the transformation. As a strong electron withdrawing group, a nitro group exhibits higher reactivity to Sn/HCl, thereby leading to selective reduction.

It will be helpful to understand these reactions by reviewing the section on the borohydride reduction of aldehydes and ketones in your organic lecture text, as well as the reduction of the nitro group (usually not covered at all in class but still described in the text).

Safety Precautions. $NaBH_4$ is harmful if inhaled or absorbed through skin. $NaBH_4$ decomposes quickly in H_2O, especially if acidic solutions are used. The decomposition forms toxic diborane and flammable/explosive hydrogen gas, and thus must be carried out in a hood. Keep the reagent bottle closed when not in use.

This experiment was adapted from A. G. Jones, *J. Chem. Educ.* **1975**, *52*, 668.

Experiment

Reduction of the Carbonyl Group—Synthesis of 1-(*m*-Nitrophenyl)ethanol

- Dissolve ~200 mg (record the exact weight) of *m*-nitroacetophenone in 2.5 mL of absolute ethanol in a 25 mL round-bottom flask.
- Warm the flask on the steam bath until all solids have dissolved. Add a magnetic stir bar and stir the solution, while cooling the flask in an ice bath to form a fine suspension of solids.
- Add approximately 50 mg (no less) of sodium borohydride to the flask and continue stirring for 5 min. (You do not need to weigh $NaBH_4$. There is a test tube at reagent bench, which contains ~50 mg of $NaBH_4$. Use that as a guide.)
- Allow the reaction to warm to room temperature and stir for 15 min after reaching room temperature.
- Add 5 mL of water to the reaction mixture.

- Extract the mixture with three 2 mL portions of methylene chloride. For each extraction, use a magnetic stirrer to thoroughly mix the layers for a few minutes then allow the layers to separate. Recover the lower layer each time via a Pasteur pipette and transfer it to a small Erlenmeyer flask.
- Dry the combined organic layers with ~0.5 g of sodium sulfate. Swirl the contents for 5 min, then recover the liquid portion with a Pasteur pipette and transfer it to 25 mL round-bottom flask.
- Carefully remove the solvent under reduced pressure.
- Keep under reduced pressure until the product solidifies.
- Recrystallize the solid from 0.5 mL of toluene.
- Collect the supernatant liquid using the Pasteur pipette, exercising care not to pipette up any solid.
- Wash solid with 0.5 mL of cold hexane and dry the crystals.
- Obtain the weight and melting point of the product. Literature melting point for the reduced product is 60–62 °C.

Reduction of the Nitro Group—Synthesis of *m*-Aminoacetophenone

- Weigh 100 mg of *m*-nitroacetophenone into a 10 mL round-bottom flask.
- Weigh approximately 0.20 g of granulated tin and add it to the contents of the flask.
- Add 2.0 mL of HCl solution to the contents of the flask.
- Equip the flask with a reflux condenser and begin to flow water through the condenser.
- Heat the flask with the heating mantle (controller = 8) until it begins to reflux. Continue to heat for 15 min after the reflux had begun. (A small amount of tin left over is OK.)
- Allow the flask to cool to room temperature.
- Pipette the reaction solution away from the unreacted tin and place in a 25 mL Erlenmeyer flask.
- Cool the flask in an ice bath.
- While the flask is cooling, obtain 1.2 mL of the premade NaOH solution in a small test tube.
- Cool the test tube in the same ice bath as the flask for a few minutes.
- Now add the cooled NaOH solution in the test tube to the reaction mixture in the flask via a Pasteur pipette dropwise with swirling. This will precipitate the product, as the free amine and convert the Sn(IV) compounds to soluble stannates.
- Stir the slurry, while cooling in an ice bath, to ensure completion of precipitation of product and dissolution of the stannates.
- Vacuum filter the solid material on the Hirsch funnel. Wash the solid twice with a minimal amount of ice water (about 1 mL). Allow to suction dry for about 10 min. You may allow the solid to dry in the Hirsch funnel until next lab and then determine the melting point.
- If you wish, you can recrystallize the product from boiling water in a small test tube. The material will crystallize out as fine yellow plates.

Laboratory Notebook

Use the format provided in the syllabus packet. You can find data on the materials used in the reaction in the Merck Index or the Aldrich catalog or similar reference material. You only need melting point data for the starting material, *m*-nitroacetophenone and the two products. Provide molecular weight data for the other materials involved. Provide the equations and the mechanisms for the two reactions.

Product

The two solid products, properly labeled.

Calculations for theoretical and percent yields, and identification of the limiting reagents for both reactions.

Clean the following glassware at the end of this lab: one 25 mL round-bottom flask, a condenser and an Erlenmeyer flask (25 mL) by first rinsing them with H_2O, then using Alconox powder which is provided at each sink. Follow the instruction posted in the lab. For removal of tough organic stain, you may use some acetone. Give the cleaned glassware to your TA, and sign the glassware check-in list. There will be a tray (or plastic bin) for the collected glassware for each session, which will be provided by the organic stock room. The collected glassware will be placed in the Organic Lab oven, and the dried glassware will be returned to you for Experiment 12.

EXPERIMENT 12

GRIGNARD REACTION

Objective

- To illustrate the utility of the Grignard reaction as well as learn to work under anhydrous conditions.

Background

Grignard reactions are C–C bond-forming reactions that produce alcohols from alkyl/aryl halides and a carbonyl group. In this reaction (eq 1), the Grignard reagent (**1**) is first prepared from an alkyl or aryl halide (I, Br, or Cl) in either diethyl ether or THF. The second step then reacts the Grignard reagent with a compound containing a carbonyl group (eq. 2) to produce the alkoxide (**2**), that is then protonated by the addition of acid (usually dilute HCl). For additional information about this important organic reaction, please read your textbook.

$$\underset{X\,=\,I,\,Br,\,Cl}{R\text{-}X} + Mg \longrightarrow \underset{\mathbf{1}}{RMgX} \qquad 1$$

$$\underset{\mathbf{1}}{RMgX} + R'C(=O)R'' \longrightarrow \underset{\mathbf{2}}{R'C(R)(OMgX)R''} \qquad 2$$

© Courtesy of David Modarelli

The mechanism for the formation of the Grignard reagent involves the transfer of two electrons from elemental Mg into the C–X bond. The resulting product is best represented as having a very polar covalent bond between carbon and magnesium, and an ionic bond between magnesium and the halide anion (i.e., **3**). Diethyl ether or tetrahydrofuran (THF) are usually used as solvents in Grignard reactions, since they can stabilize the Grignard reagent by complexing the Mg^{2+}(**4**).

$$\underset{\mathbf{3}}{R^{-}Mg^{2+}X^{-}} \qquad \underset{Et_2O}{Et\text{–}O\text{–}Et} \qquad \text{THF} \qquad \underset{\mathbf{4}}{R^{-}Mg^{2+}X^{-}(\ddot{O}Et_2)_2}$$

© Courtesy of David Modarelli

The carbanion portion of the Grignard reagent is very basic (recall that the pK_a values of most hydrocarbons are in the 45–50 range) as a result of the polar carbon–magnesium bond, and therefore reacts rapidly with any acidic protons present in solution. For this reason, the halide used to prepare the Grignard reagent, the reacting carbonyl group, and the solvent cannot contain acidic protons. Water (pK_a 15.7) is often found in organic solvents in small amounts and will immediately react with and therefore destroys a Grignard reagent. To avoid the reaction between a Grignard reagent and water, the glassware and the ether solvent are thoroughly dried prior to use. In this lab, we will use pre-dried diethyl ether (**NOTE: we will all use the same bottles of diethyl ether, and so you should screw the cap back on tightly after you have taken what you need for your reaction!**) and will use oven-dried glassware to minimize the presence of water. If you turned in your glassware last lab to the stockroom for drying, collect an oven-dried CLEAN 25 mL round-bottom flask, a condenser and an Erlenmeyer flask from the stockroom for the experiment.

The Grignard reaction in this lab will be between phenyl magnesium bromide and benzophenone (Scheme 1). In this reaction, we will first form the Grignard reagent (eq 3) by reacting strips of magnesium (called magnesium turnings) with bromobenzene in dry diethyl ether. This reaction can be slow to start, and so you will warm your solution to reflux using a heating mantle. The Grignard reagent has formed when your solution turns from clear to light brown or gray. Once the reaction is complete you will cool the solution to room temperature before continuing to the second step.

In the second step (eq 4), you will slowly add (using a pipette) an ether solution of benzophenone to the stirred solution of the Grignard reagent through the top of the reflux condenser. The reaction between the two compounds is exothermic and will cause the ether to reflux—be sure to add the benzophenone slowly enough so that the solution does not reflux too vigorously. You will then reflux for an additional 10 min using a heating mantle.

Br $\xrightarrow[Et_2O]{Mg}$ MgBr 3

MgBr + O $\xrightarrow{Et_2O}$ $O^-Mg^{2+}Br^-$ 4

$O^-Mg^{2+}Br^-$ $\xrightarrow[H_2O]{HCl}$ OH 5

Experiment

IMPORTANT: ***Remember water (in air, bench tops, and glassware) will destroy the Grignard reagent. Do not use ANY wet glassware or allow the reaction to come into contact with water. Close the ether can after every use. Do not be "polite" and leave it open for the next person, as you may have just ruined their experiment.***

Synthesis of Phenylmagnesium Bromide

- Weigh out 180 ± 20 mg of magnesium turnings.
- Put the turnings in a clean, oven-dried 25 mL round-bottomed flask.
- Add a magnetic stir bar and an oven-dried reflux condenser to the flask. The reflux condenser should not be removed until after the HCl addition (much later). Place the flask in heating mantle over a magnetic stirrer.

- Add ~6 mL of anhydrous ether using a clean, dry syringe/pipet into an oven-dried Erlenmeyer flask and then add it to the already oven-dried 25 mL round-bottomed flask that contains your stir bar and condenser.
- Turn on the magnetic stirrer.
- Using a syringe/pipet, measure out 0.75 mL of bromobenzene and add it in one portion through the top of the reflux condenser.
- Set the heating mantle controller to 2.
- Reflux the mixture for 15 min after the solution starts to turn brown.
- If the solution does not turn brown, remove the round bottom flask from the condenser and CAREFULLY grind the magnesium turnings with a stir rod to initiate the reaction. The solution should turn cloudy and slowly change to a brown color if the Grignard reagent formation is initiated.
- Remove the heating mantle and cool the reaction to room temperature.

Synthesis of Triphenylmethanol

- Dissolve 0.911 g of benzophenone in 6 mL of ether in a dry 25 mL Erlenmeyer flask. Add this solution dropwise with a syringe/pipette through the top of the condenser at such a rate as to maintain a gentle reflux. Try to keep the stirring going during the addition.
- After the addition is complete; reflux the solution for 10 min. Remove the heating mantle and cool the reaction in an ice bath.
- Add 5 mL of 6 M HCl in 0.5 mL increments over a 5 min period.
- Remove the ice bath and stir at room temperature until the salts dissolve.
- Transfer the solution to a separatory funnel and separate the layers. Dry the ether layer in a 50 mL Erlenmeyer flask over Na_2SO_4. Separate the ether from the drying agent and place in a 50 mL Erlenmeyer flask. Evaporate the ether carefully on the steam bath.
- After the ether has evaporated, cool the flask to room temperature, add 10 mL of hexane, mix well, and break up the solid. Cool the hexane solution in an ice bath.
- Collect the crystals using a Hirsch funnel and then wash them with 5 mL of cold hexane. Draw air through the crystals (while still on the Hirsch funnel) for a few minutes to dry the product.

Laboratory Notebook

- As per the syllabus.
- Full prelab.

Product

- Triphenylmethanol and its melting point.
- Calculations of limiting reactant, theoretical yield, and percent yield for the reaction.

EXPERIMENT 13

INFRARED SPECTROSCOPY

Objective

- To demonstrate the use of infrared spectroscopy (IR) to characterize organic compounds.

Reading Assignment

- *Your Organic Text:* Chapter on Infrared Spectroscopy.

Experiment

You will be, with assistance from your TA, obtaining an infrared spectrum of an unknown compound that you will obtain from the stockroom. Then analyze the data to determine the most likely structure for your unknown.

Step I: Obtaining the spectrum.

- To a NaCl plate, apply one drop of your unknown sample (***please keep these plates away from H_2O and wear gloves when handling them!***). Place the second NaCl plate onto the first such that the sample becomes a thin film pressed between the two plates.
- Your TA will assist you in the utilization of the software to obtain your spectrum.
- After acquiring the spectrum, clean the NaCl plate by washing off the unknown with CH_2Cl_2.

Step II: Obtain the molecular formula of the unknown and perform a HDI calculation.

- On the worksheet provided, enter your unknown number and the molecular formula provided to you by your TA.
- On the worksheet perform the HDI calculation given to you in class to determine the number of multiple bonds and/or rings your molecule contains.

Step III: Identify the functional groups in your molecule using the IR spectrum.

- On the worksheet, list the peak frequencies that you observe on your spectrum. Enter in using the data tables provided to you or in your book the functional groups corresponding to frequencies. You will need to review the chapter of Infrared spectroscopy in your textbook, for the peak frequencies of different functional groups.

- Keep in mind that there are no halogens, (F, C1, Br, or I), no sulfur, no phosphorus, no nitro or nitroso-groups, and no carboxylic acids, acid halides, anhydrides, or aromatic heterocylces in any of the unknowns. You are not responsible for the substitution pattern on any aromatic rings that you encounter in the unknowns, simply list the groups you feel are attached to the ring.

Step IV: Identify your Unknown.

Using the HDI (which gives you the combinations of rings, double or triple bonds possible), the molecular formula (the combination of elements possible), and the functional groups present and being to construct possible structures for you unknown. Enter the best structure of your unknown in the box provided on the worksheet.

Laboratory Notebook

- Signature of TA with date of IR spectrum.
- No prelab required.
- Completed worksheet.

Product

- Identification of unknown.

Organic Laboratory I Name ______________________________

Experiment 13: Infrared Spectroscopy Worksheet

Unknown Number : ______________________

Molecular Formula of Unknown : ______________________

Calculation of the Hydrogen Deficiency Index (HDI): Using the equation given to you in class determine the number of combinations of rings and/or double or triple bonds in your unknown.

HDI = ____________

Possible combinations of rings and/or double and triple bonds (use *r* for rings, *db* for double bonds, *tb* for triple bonds):

Identification of peaks in the IR spectrum:

Frequency cm^{-1}	Possible functional group

Identification of Unknown:

Structure:

Name of compound:

www.ingramcontent.com/pod-product-compliance
Ingram Content Group UK Ltd.
Pitfield, Milton Keynes, MK11 3LW, UK
UKHW051136260726
13967UKWH00010B/3091

9 781792 438240